Mc
Graw
Hill
Education

AF838640

McGraw-Hill

My Math

Grade K

Think Smart for the Smarter Balanced Assessment

www.mheonline.com

McGraw Hill Education

STEM

McGraw-Hill is committed to providing Instruction materials in Science, Technology, Engineering, and Mathematics (STEM) that give all students a solid foundation, one that prepares them for college and careers in the 21st century.

Send all inquiries to:
McGraw-Hill Education
8787 Orion Place
Columbus, OH 43240

ISBN: 978-0-02-132062-2
MHID: 0-02-132062-4

Printed in the United States of America.

2 3 4 5 6 7 8 9 10 RHR 20 19 18 17 16 15

STEM

Our mission is to provide educational resources that enable students to become the problem solvers of the 21st century and inspire them to explore careers within Science, Technology, Engineering, and Mathematics (STEM) related fields

Contents

Teacher's Guide to Smarter Balanced Assessment Preparation

Whether it is the print ***Think Smart for Smarter Balanced Assessment*** or online at **ConnectED**.mcgraw-hill.com, *McGraw-Hill My Math* helps students prepare for the SBAC testing.

How to Use this Book

Think Smart for the Smarter Balanced Assessment includes experiences needed to prepare students for the upcoming SBAC assessment. The exercises in this book give students a taste of the different types of questions that may appear on the SBAC test.

Smarter Balanced Assessment Item Types

- Familiarizes students with commonly-seen item types
- Each type comes with a description of the online experience, helpful, hints, and a problem for students to try on their own.

Countdown to SBAC

- Prepares students in the 20 weeks leading up to the SBAC assessment
- Consists of five problems per week, paced with order of the *McGraw-Hill My Math Student Edition* with built in review.
- **Ideas for Use** Begin use in October for pacing up to the beginning of March. Assign each weekly countdown as in-class work for small groups, homework, a practice assessment, or a weekly quiz. You may assign one problem per day or have students complete all five problems at once.

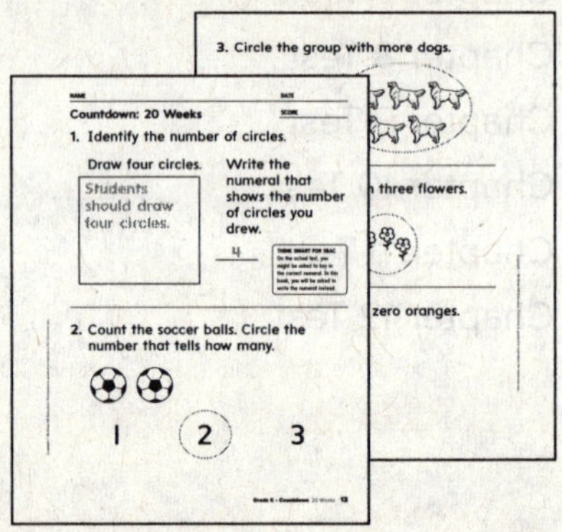

Chapter Tests

- Each six-page test assesses all of the Common Core State Standards for Mathematics presented in the chapter.
- Each question mirrors an item type that might be found on the SBAC assessment, including multi-part questions.
- **Ideas for Use** Assign as in-class group work, homework, a practice assessment, a diagnostic assessment before beginning the chapter, or a summative assessment upon completing the chapter.

Chapter Performance Tasks

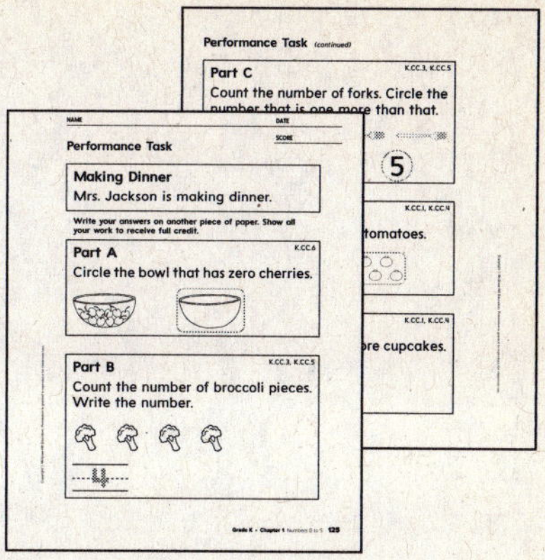

- Each two-page performance task measures students' abilities to integrate knowledge and skills across multiple standards. This helps students prepare for the rigor expected in college and future careers.
- A rubric describes the standards assessed and guidelines for scoring student work for full and partial credit.
- Sample student work is also included in the answer section of this book.
- **Ideas for Use** Assign as in-class small group work, homework, a practice assessment, or in conjunction with the Chapter Test as part of the summative assessment upon completion of the chapter.

Benchmark Tests

Four benchmark tests are included in this book. All problems on the tests mirror the item types that may be found on the SBAC assessment. Each benchmark test also includes a performance task.

- The ***first*** benchmark test is an eight-page assessment that addresses the CCSS from the first third of the Student Edition.
- The ***second*** benchmark test addresses the second third of the Student Edition.
- The ***third*** and ***fourth*** benchmark tests (Forms A and B) are twelve-page assessments that address the CCSS from the entire year, all chapters of the Student Edition.
- A rubric is provided in the Answer section for scoring the performance task portion of each test.
- **Ideas for Use** Each benchmark test can be used as a diagnostic assessment prior to instruction or as a summative assessment upon completion of instruction. Forms A and B can be used as a pretest at the beginning of the year and then as a posttest at the end of the year to measure mastery progress.

Go Online for More! connectED.mcgraw-hill.com

Additional question analysis for each item in this book is available online. The analysis includes SBAC claims, DOK levels, CCSS assessed, Mathematical Practices implemented, and more.

Performance Task rubrics to help students guide their responses are also available. These describe the tasks students should perform correctly in order to receive maximum credit.

Additional year-end performance tasks are available for Grades 1 through 5 in ***Countdown to Common Core*** blackline masters available under Assessment in ConnectED.

Students can also be assigned tech-enhanced questions from the eAssessment Suite in ConnectED. These questions provide not only rigor, but the functionality students may experience when taking the online SBAC assessment.

Smarter Balanced Assessment Item Types

In Grade 3, students will begin to take the Smarter Balanced Test for Math. It is a test that is taken on a computer. The problems on the next few pages show the kinds of questions students might have to answer and what to do to show the answer on the computer. However, the student materials in this book center mainly on ***Multiple Choice*** and ***Click to Select*** formats.

Selected Response means that you are given answers from which you can choose.

Selected Response Items

Regular multiple choice questions are like tests you may have taken before. Read the question and then choose the one best answer.

Multiple Choice

Sasha has a collection of coins. There are 3 in one pile and 2 in another. How many coins does she have in all?

- ☐ 2
- ☐ 3
- ☐ 5

ONLINE EXPERIENCE Click on the box to select the one correct answer.

HELPFUL HINT Only one answer is correct. You may be able to rule out some of the answer choices because they are unreasonable.

Sometimes a multiple choice question may have more than one answer that is correct. The question may or may not tell you how many to choose.

Multiple Correct Answers

Choose **all** the statements that tell the story.

- ☐ $2 + 3 = 5$
- ☐ $5 = 2 + 3$
- ☐ $10 - 2 = 8$

ONLINE EXPERIENCE Click on the box to select it.

HELPFUL HINT Read each answer choice carefully. There may be more than one right answer.

Another type of question asks you to tell whether the sentence given is true or false. It may also ask you whether you agree with the statement, or if it is true. Then you select yes or no to tell whether you agree.

Multiple True/False or Multiple Yes/No

There were no birds on the pond. Two brown ducks swim into the pond. Five geese fly in and land on the pond. Tell whether you agree with each sentence.

Yes	No	
☐	☐	There are more geese than ducks on the pond.
☐	☐	There are 2 birds on the pond.
☐	☐	5 geese flew away.
☐	☐	There are 7 birds in all on the pond.

ONLINE EXPERIENCE Click on the box to select it.

HELPFUL HINT There is more than one statement. Any or all of them may be correct.

You may have to choose your answer from a group of objects.

Click to Select

Which shape is like a car tire?

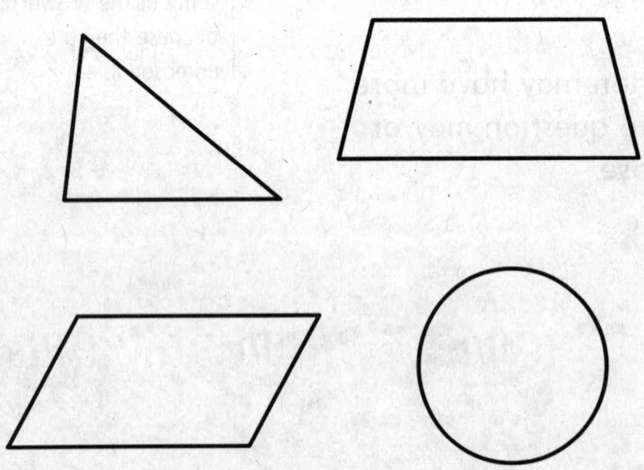

ONLINE EXPERIENCE Click on the figure to select it.

HELPFUL HINT On this page you can draw a circle or a box around the figure you want to choose.

Sometimes you must use your mouse to click on an object and drag it to the correct place to create your answer.

Drag and Drop

Move these objects so that they match the story.

ONLINE EXPERIENCE You will click on an object and drag it to the spot it belongs.

HELPFUL HINT Either draw a line to show where the object goes or sketch the object in the right place.

The bear sits on the chair.

The elephant sits on a shelf above the chair.

The giraffe sits below the chair.

When no choices are given from which you can choose, you must create the correct answer. One way is to type in the correct answer. Another may be to make the correct answer from parts that are given to you.

Constructed-Response Items

Fill in the Blank

Loren has 3 fish in her tank. For her birthday she will receive 2 more fish. Write a number sentence to show how many fish she will have after her birthday.

ONLINE EXPERIENCE You will click on the space and a keyboard will appear for you to use to write the numbers and symbols you need.

HELPFUL HINT There is more than one way to write the correct sentence.

_______ ○ _______ = _______

NAME DATE

Countdown: 20 Weeks

SCORE

1. Identify the number of circles. K.CC.3

Draw four circles.

Write the numeral that shows the number of circles you drew.

2. Count the soccer balls. Circle the number that tells how many. K.CC.4a

1 2 3

3. Circle the group with more dogs. K.CC.4a

4. Circle the group with three flowers. K.CC.4a

5. Circle the bowl with zero oranges. K.CC.4a

NAME DATE

Countdown: 19 Weeks

SCORE

1. Write the number of umbrellas shown. K.CC.3

2. Count the spoons. Find the group of forks with the same number. Draw a line from the group of spoons to the group of forks with the same number. K.CC.4a

3. Draw an X on the groups that show less than 4. K.CC.4a

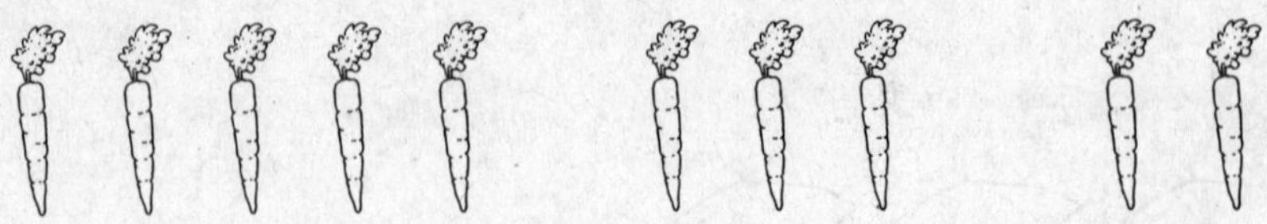

4. Circle the number that is one more than two. K.CC.6

2   3   1

5. Draw a group of bugs that is less than the group shown. K.CC.4a

NAME .... DATE ....

Countdown: 18 Weeks

SCORE ....

1. Count and draw lines to match objects in one group with objects in the other. Write the numbers. Circle the number and group that is greater than the other. K.CC.3

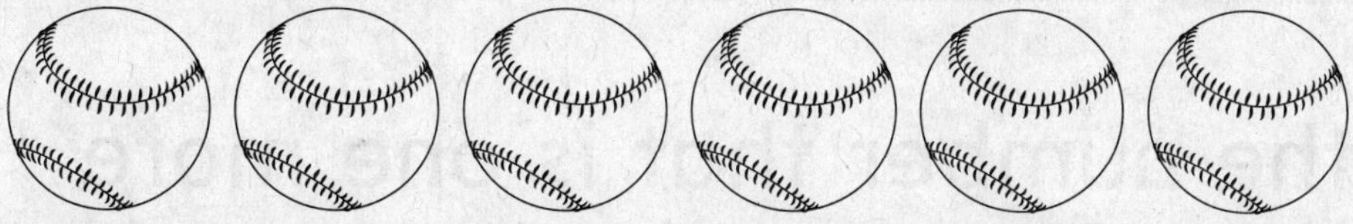

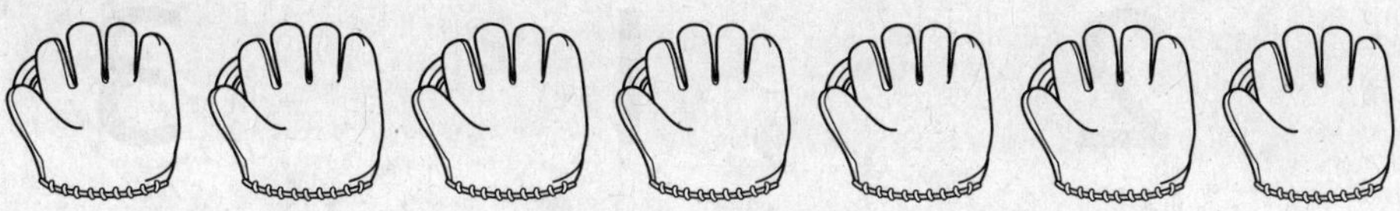

2. Draw a box around the second rock from the frog. Draw an X through the third rock. Draw a line under the first rock. K.CC.4a

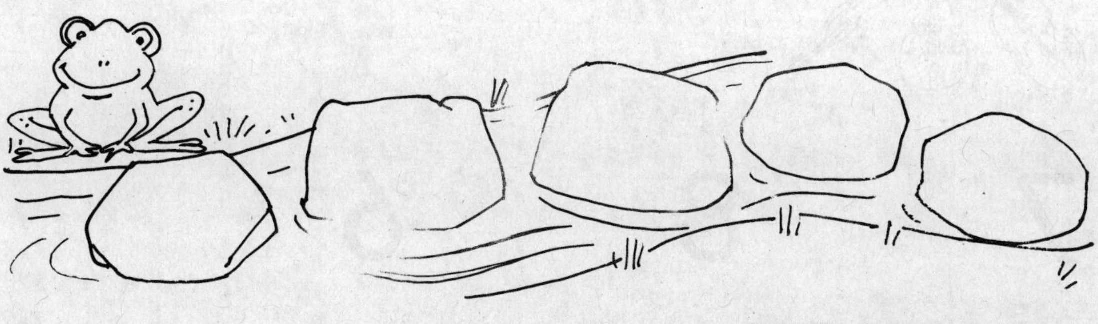

3. Count the houses. Draw one more house. Count how many. Write the number. K.CC.3

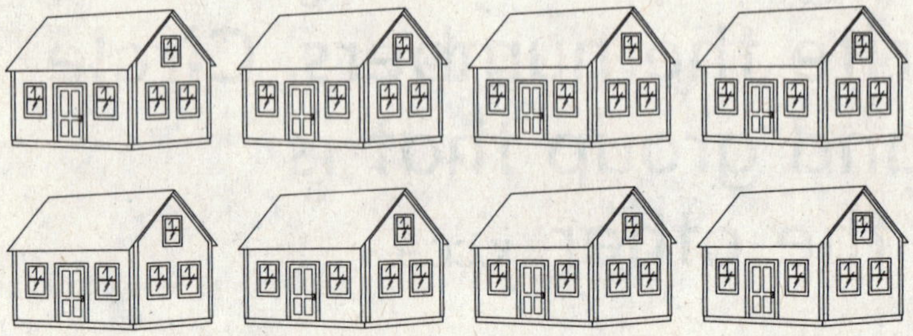

4. Circle the number that is one more than one. K.CC.4c

0 2 1 3

5. Count the violins. Circle the number that tells how many. K.CC.4b

7 8 6

NAME DATE

Countdown: 17 Weeks

SCORE

1. Circle the group with four soccer balls. K.CC.4a

2. Count the number of animals. Circle the number that is one more than that. K.CC.4c

5 4 3

3. Color the boxes in the ten-frame to show eight. K.CC.1

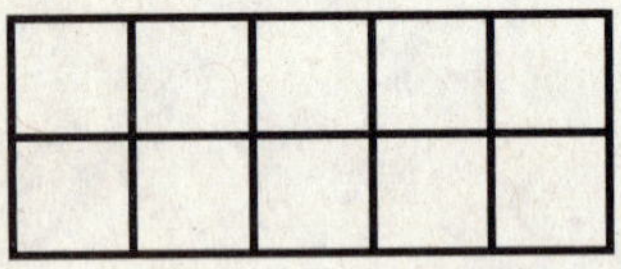

4. Circle the number that is one more than two. K.CC.4c

0 2 1 3

5. Count the pumpkins. Write the number that tells how many. K.CC.3

NAME DATE

Countdown: 16 Weeks

SCORE

1. Count the giraffes. Write how many. K.CC.3

2. Circle 15 geese. K.CC.3

3. Draw more circles to make 19. K.CC.3

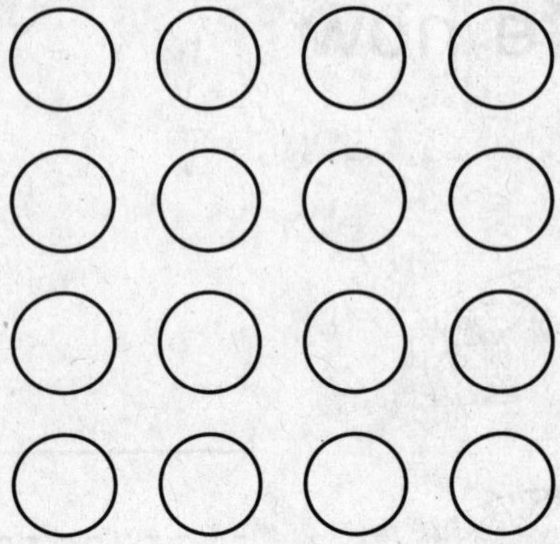

4. Connect the numbers in order. K.CC.3

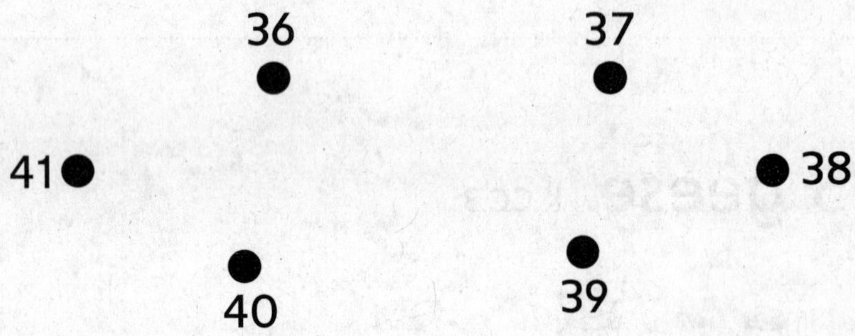

5. Circle the number that is one more than 1. K.CC.4c

1 0 2

NAME ______________________ DATE ______________

Countdown: 15 Weeks

SCORE ______________

1. Count the pieces of pasta. Write one more than the number. K.CC.3

---

2. Count and draw lines to match objects in one group with objects in the other. Write the numbers. Circle the group that has more than the other. K.CC.3

3. Color more boxes in the ten-frame to show 18. K.CC.1

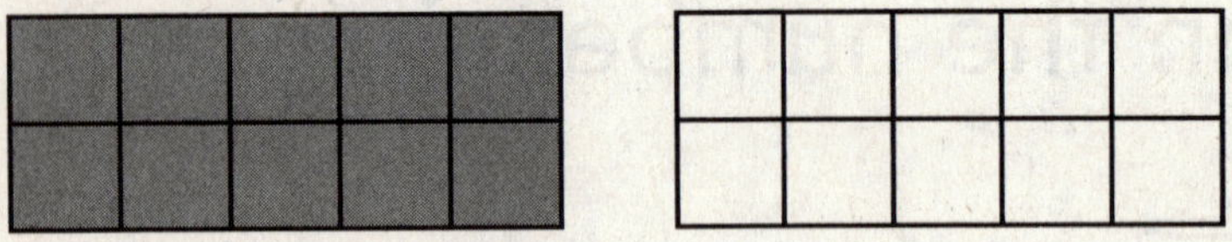

4. Count the presents. Circle the number that is one more than that. K.CC.4c

2 3 4

5. Circle the fourth strawberry from the basket. K.CC.4a

NAME DATE

Countdown: 14 Weeks

SCORE

1. Count by tens. Write the number. K.CC.3

2. There are 4 cars. Take apart the number. K.OA.1

and

3. Count how many. Write the number beside the carrots. Take apart the number. K.OA.3

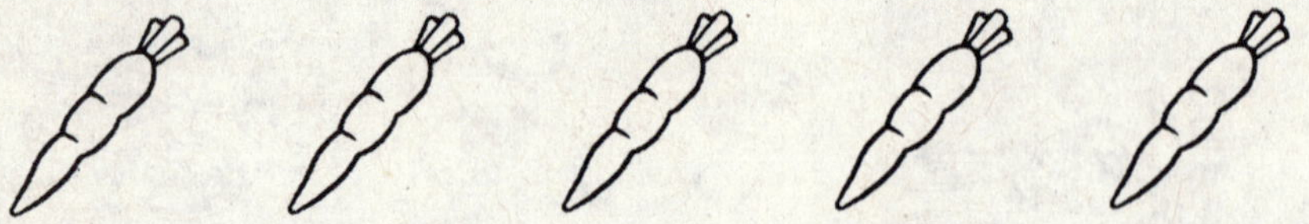

1 and

4. There are 6 cats. Circle cats to make 4 cats and 2 cats. K.OA.7

5. There are 7 chipmunks. Take apart the number. K.OA.1

and

NAME DATE

Countdown: 13 Weeks

SCORE

1. Count the balloons. Write the number beside the balloons. K.OA.1

______

Show a way to take apart the number.

______ and ______

2. There are 9 squares. Break apart a group of 6. Write the number in the other group. K.OA.3

9

6 and ______

3. Take apart 10. Write the numbers that make 10. K.OA.4

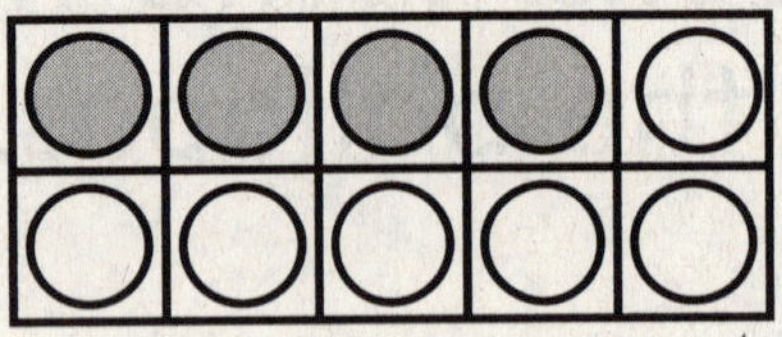

__________ and __________

4. Continue counting by tens. K.CC.3

60 __________ __________ __________

5. Circle the group with 6 pineapples. K.CC.3

NAME DATE

Countdown: 12 Weeks

SCORE

1. There are 10 crabs. Take apart the crabs into 2 groups. Write the numbers. K.OA.3

________ and ________

2. There are two crayons on the table. Joey puts two more crayons on the table. How many crayons are there in all? K.OA.1

____

3. Write the number that tells how many there are in all. K.OA.1

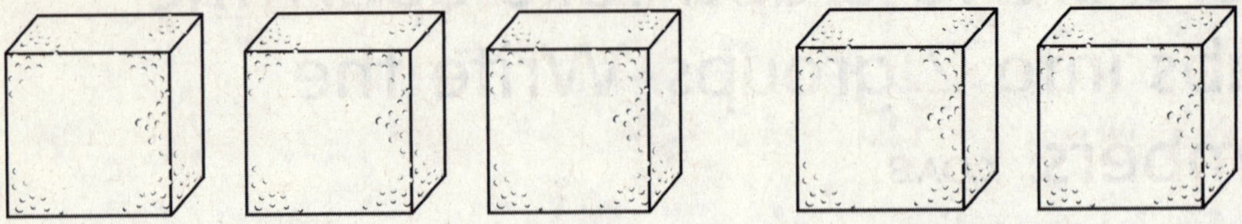

3 + 2 = ______ in all

4. Count the number in each group. Write the number. Write how many in all. K.OA.1

______ + ______ = ______ in all

5. Circle two groups to match the number sentence. K.OA.1

$2 + 5 = 7$

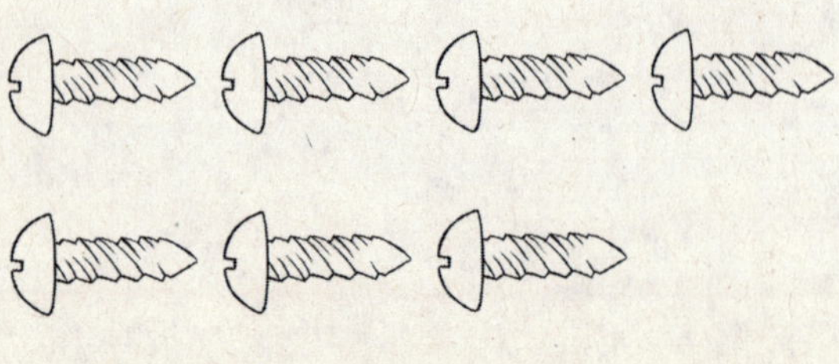

NAME DATE

Countdown: 11 Weeks

SCORE

1. There are 10 hats. Take apart the hats into 2 groups. Write the numbers. K.OA.3

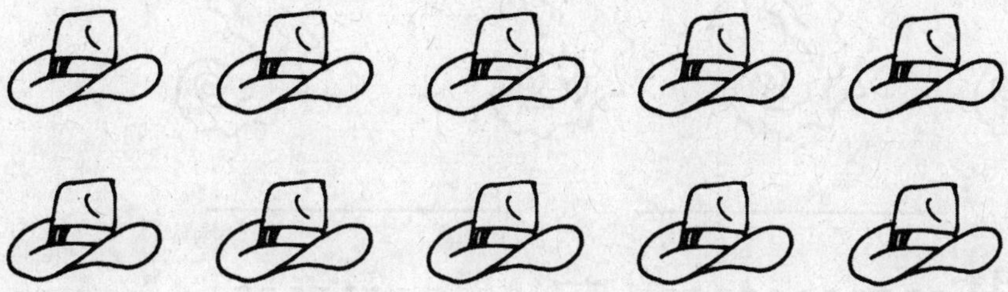

______ and ______

2. Count the number in each group. Write the numbers. Then write the number in all. K.OA.1

______ + ______ = ______

3. There are 4 seahorses. 1 more comes. How many seahorses are there in all? Write a number sentence. K.OA.1

_____ + _____ = _____

4. Count the objects and write the numbers. K.OA.4

_____ + _____ = _____

5. Count the cows. Write one more than that many. K.CC.3

_____

NAME DATE

Countdown: 10 Weeks

SCORE

1. Count the number in each group. Write the numbers. Then write the number in all. K.OA.1

_____ + _____ = _____

2. Kumar had 3 shovels. He took one away. Cross out one shovel. Write how many are left. K.OA.1

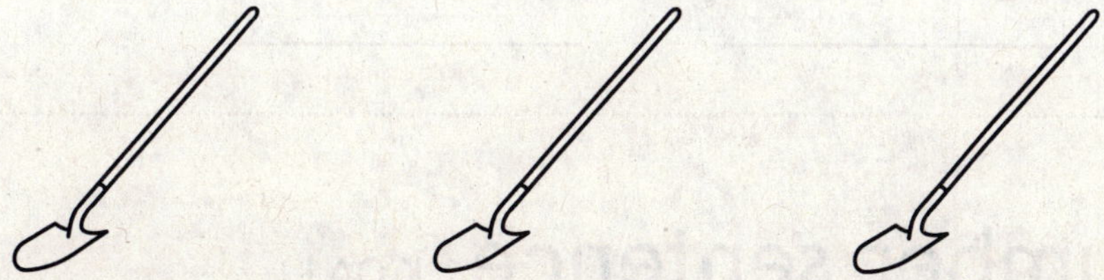

_____

3. There are 7 pieces of pizza. Enrique eats 2 pieces. Cross out two pieces. Write how many pieces are left. K.OA.1

______

4. Count the number of snakes. Count how many are going away. Count how many are left. Write the numbers. K.OA.1

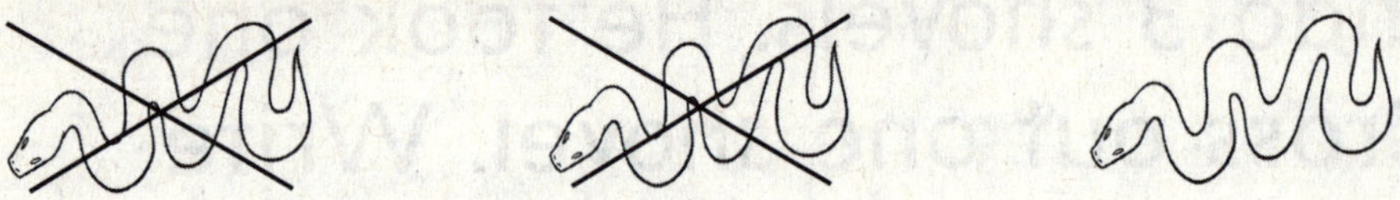

______ – ______ = ______

5. Write a number sentence. K.OA.1

______ – ______ = ______

NAME

DATE

Countdown: 9 Weeks

SCORE

1. Count the soccer balls. Three are kicked away. Cross out three soccer balls. Write how many are left. K.OA.1

---

2. There are 8 spiders. Four spiders crawl away. How many are left? Write a number sentence. K.OA.1

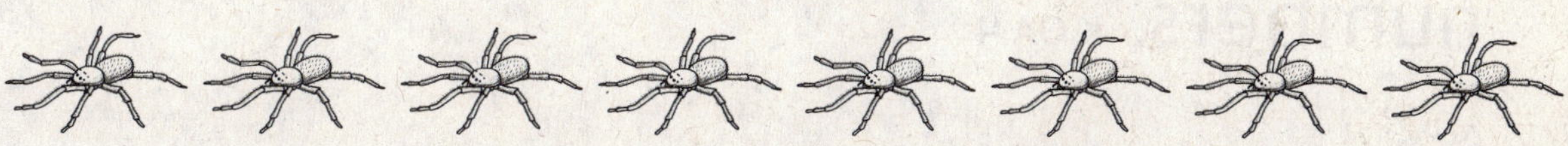

____ – ____ = ____

3. There are ten starfish. Eight starfish go back in the ocean. Cross out eight starfish. Write how many starfish are left. K.OA.3

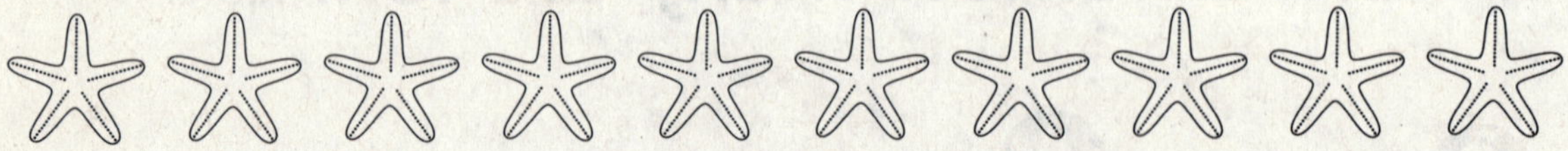

4. Write a number sentence. K.OA.1

____ + ____ = ____

5. Count the objects and write the numbers. K.OA.4

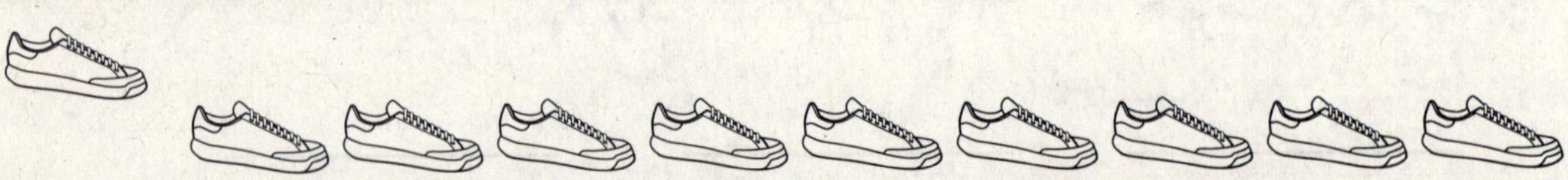

____ + ____ = 10

NAME ______________________ DATE ______________

Countdown: 8 Weeks

SCORE ______________

1. Draw an X through the strawberries to show how many you would take away. K.OA.7

$$6 - 4 = 2$$

2. Count the numbers in each ten-frame. Write the numbers. Write how many in all. K.NBT.1

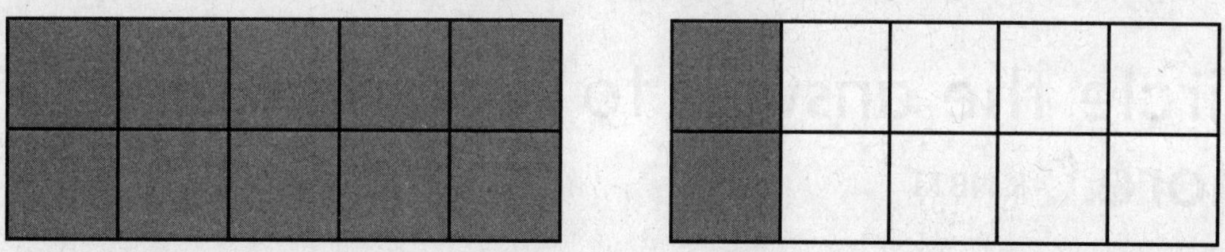

__________ and __________ is __________

3. Shade the ten-frames to show "10 and 3 is 13." K.NBT.1

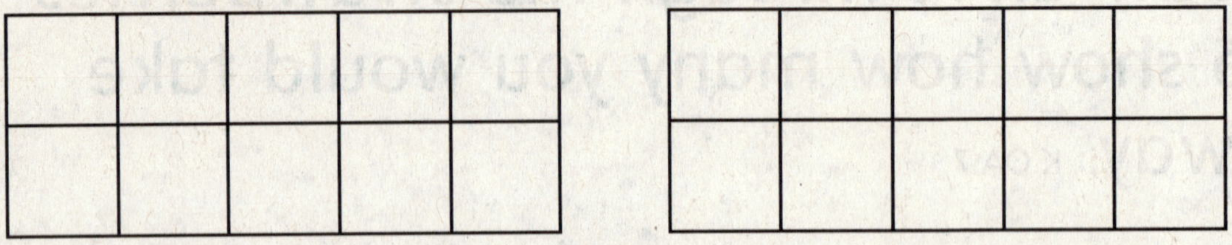

4. Count the number of each present. Write each number. K.NBT.1

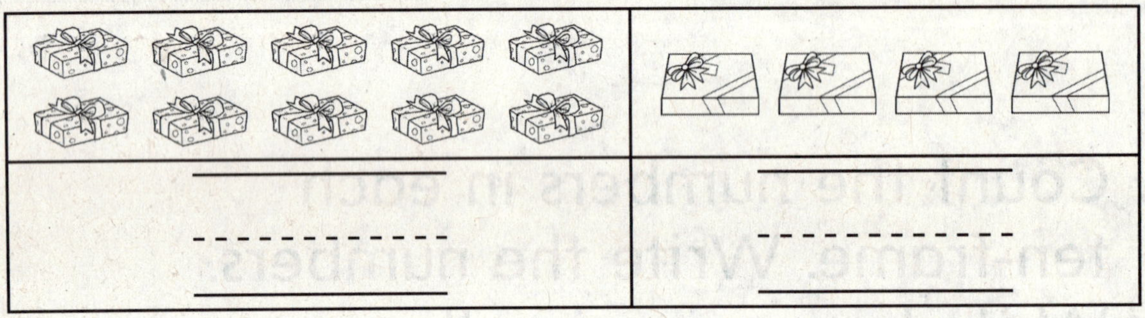

5. Circle the answer to "ten and one more." K.NBT.1

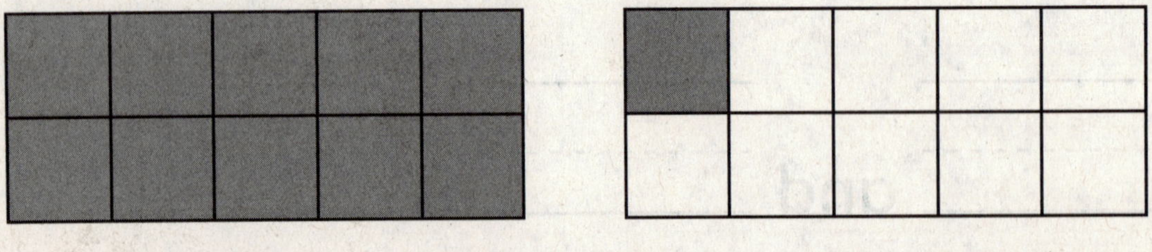

10 11 12

NAME ____________________ DATE ____________________

Countdown: 7 Weeks

SCORE ____________________

1. Match each word to the correct symbol. K.OA.1

minus sign	+
plus sign	−
equal sign	=

2. Circle the picture that shows 10 − 3. K.OA.3

3. Shade the ten-frames to show "10 and 5 more." Write the total number. K.NBT.1

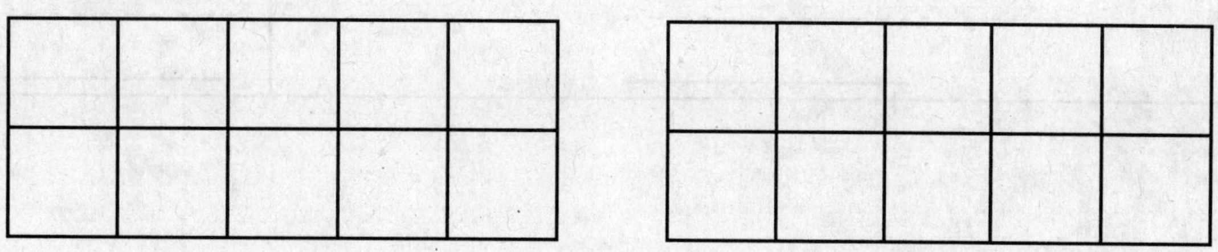

____

4. Which is correct? K.NBT.1

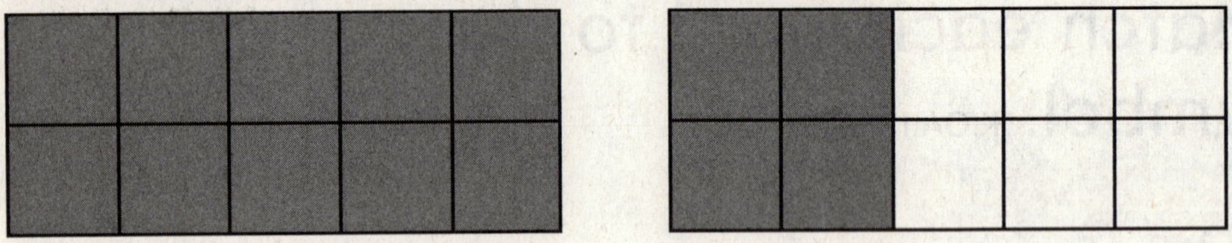

10 and 2 more

10 and 3 more

10 and 4 more

---

5. Count the number of each animal. Write each number. K.NBT.1

NAME DATE

Countdown: 6 Weeks

SCORE

1. Count the numbers in each ten-frame. Write the numbers. Write how many in all. K.NBT.1

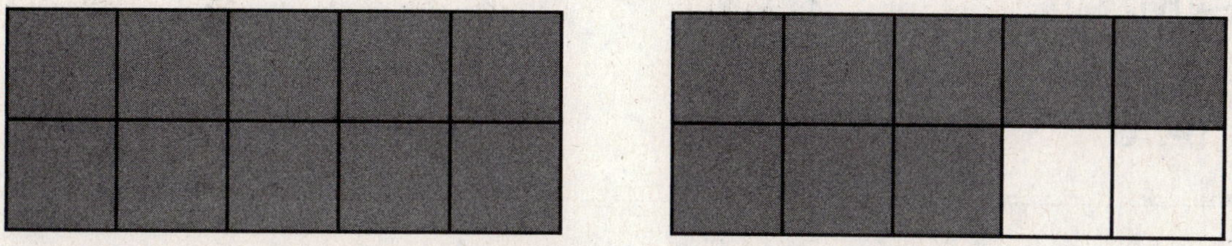

______ ______ ______

______ and ______ is ______

2. Compare the objects. Circle the one that is longer. Put an X through the one that is shorter. K.MD.2

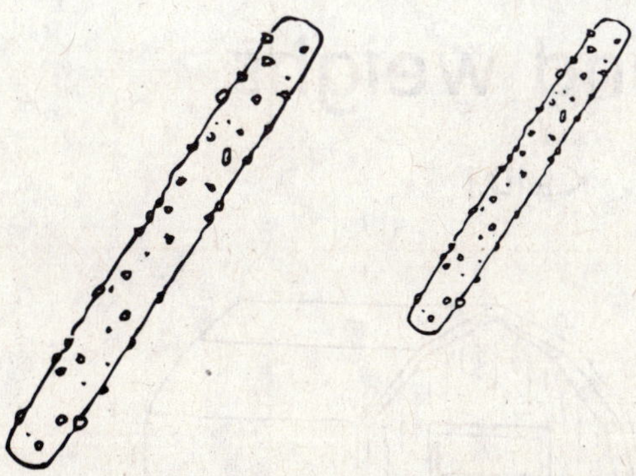

3. Compare the giraffes. Circle the taller one. Underline the shorter one. K.MD.2

4. Circle the things that are too heavy to lift. K.MD.2

5. Look at the two objects. Circle the object that is taller and weighs more than the barrel. K.MD.1

NAME ________________ DATE ________________

Countdown: 5 Weeks

SCORE ________________

1. Circle the ones that are alike. K.MD.3

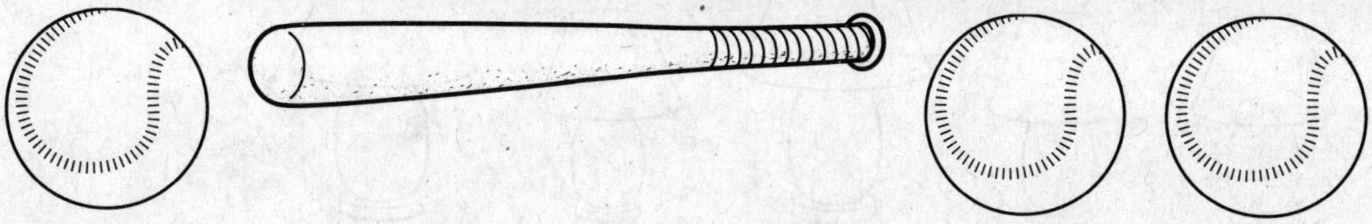

2. Look at the pictures. Circle the pictures that do not belong in a house. K.MD.3

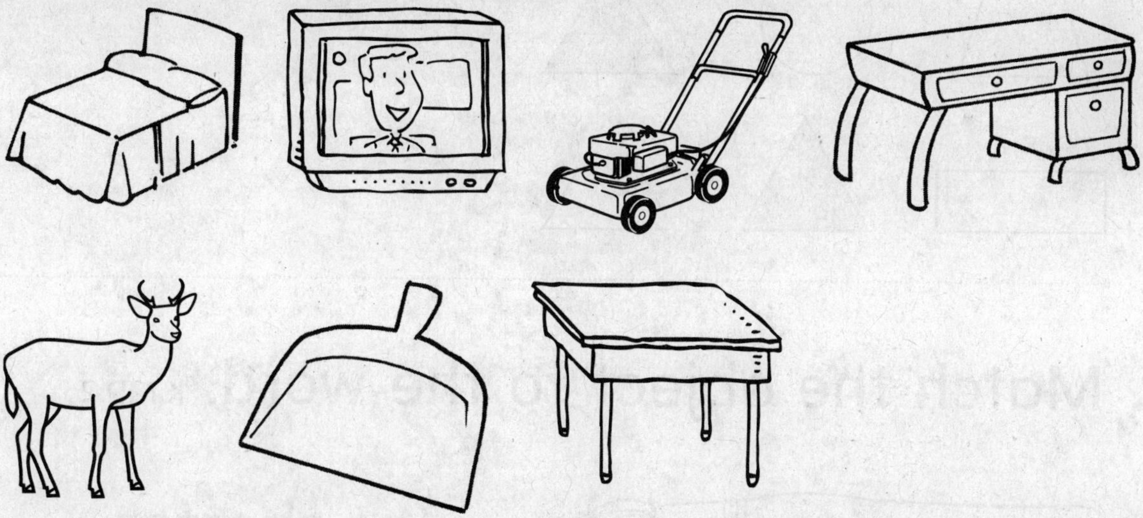

3. Sort by size. Draw X's through the bigger ones. K.MD.3

4. Sort by shape. Circle all the ones that are alike. K.MD.3

5. Match the object to the word. K.MD.2

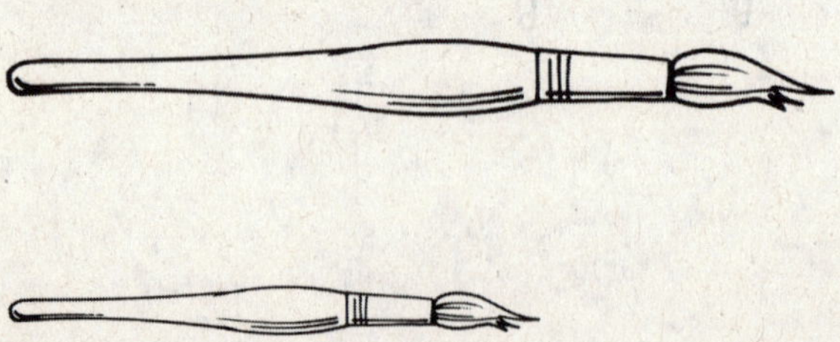

shorter

longer

NAME DATE

Countdown: 4 Weeks

SCORE

1. Circle the heavier object. Cross out the lighter object. K.MD.2

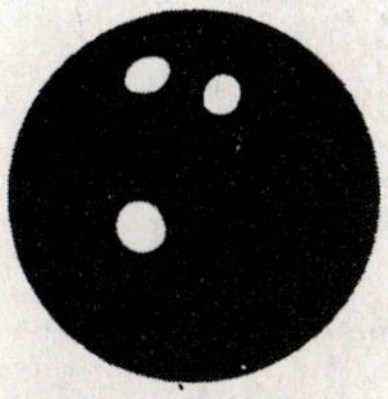

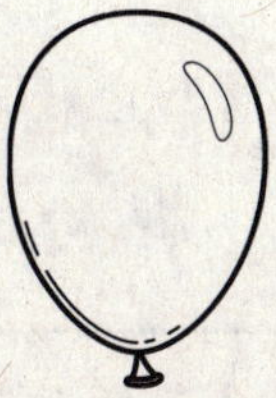

2. Circle the object that is above the log. K.G.1

3. Circle the dog that is in front of the doghouse. K.G.1

4. Circle the animals that are next to the duck. K.G.1

5. Draw a circle beside the key. K.OA.7

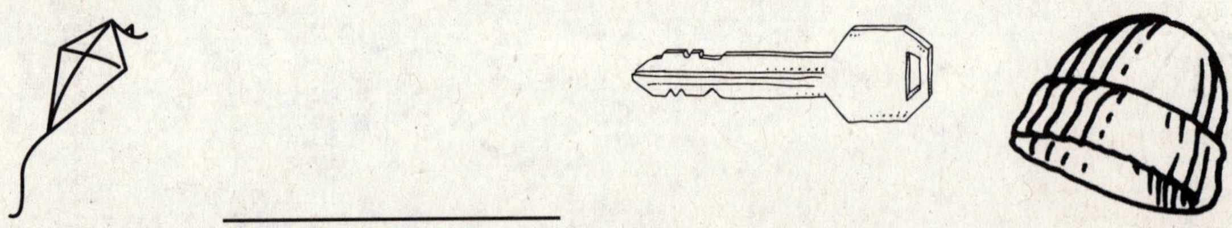

NAME .... DATE ....

Countdown: 3 Weeks

SCORE ....

1. Put an X on the squares. K.G.2

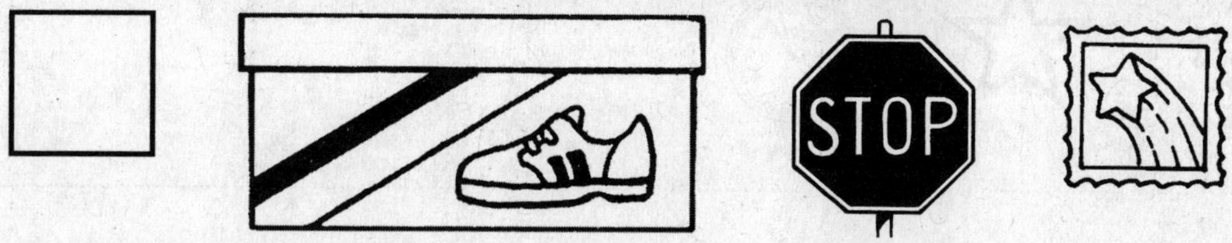

2. Match the shape to the name. K.G.4

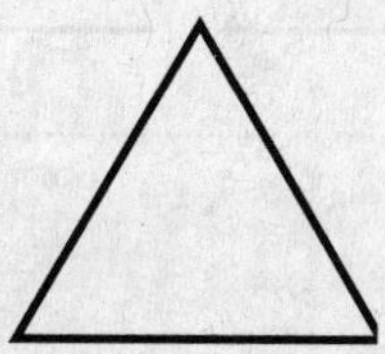

circle

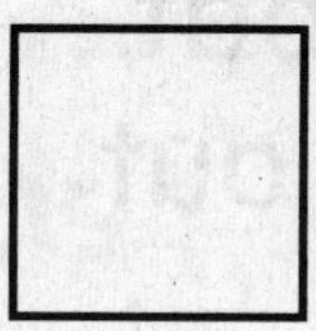

square

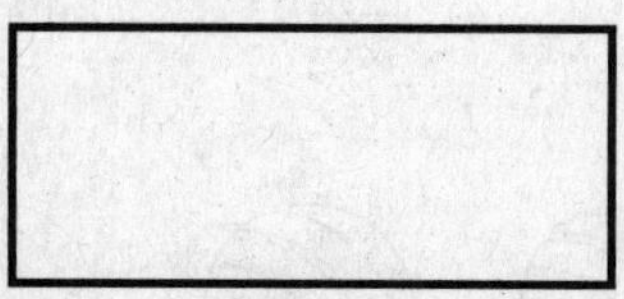

triangle

rectangle

3. Underline the object that has a hexagon. K.G.2

4. Draw the shape that comes next. K.G.4

 ______

5. Circle the fifth ball from the bat. Underline the first ball. Cross out the second ball. K.CC.4a

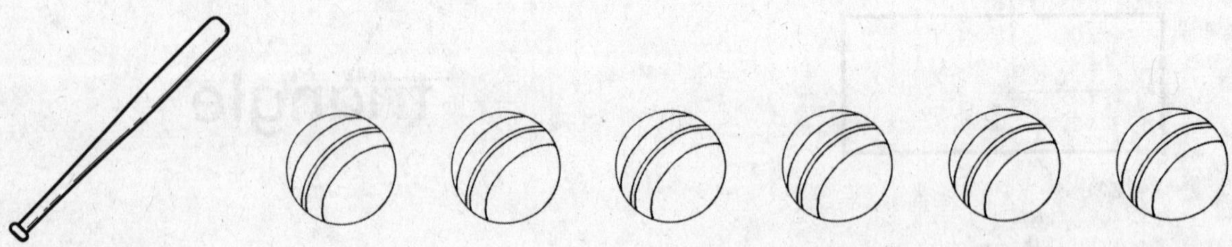

NAME .... DATE ....

Countdown: 2 Weeks

SCORE ....

1. Circle the one that is different. K.MD.3

2. Circle the shapes that are used to make the design. K.G.5

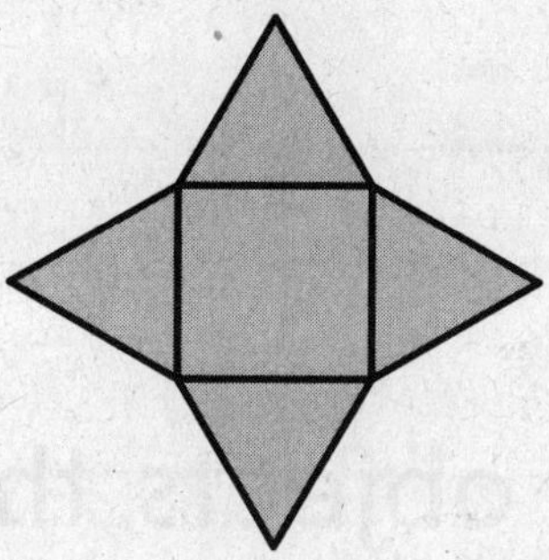

3. Draw the shape that comes next. K.G.4

______

4. Cross out the pictures that do not have circles. K.G.2

5. Underline the two objects that hold the same amount. K.MD.2

NAME DATE

Countdown: I Week

SCORE

1. Put an X on the shapes that are not spheres. K.G.2

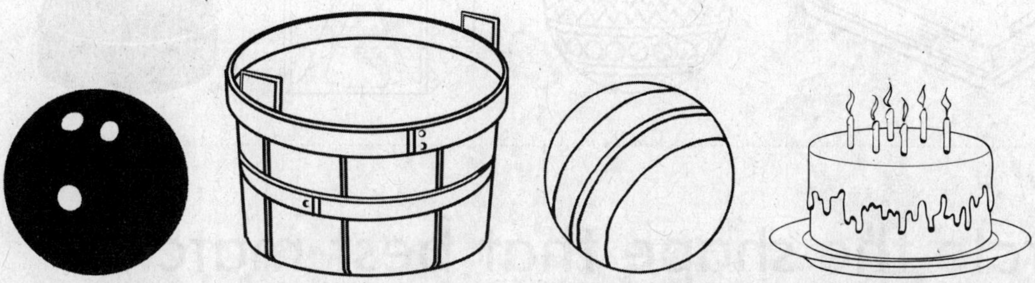

2. Circle the cone. K.G.I

3. Underline the shapes that can be stacked. K.G.4

4. Circle the shape that best matches the drum. K.OA.7

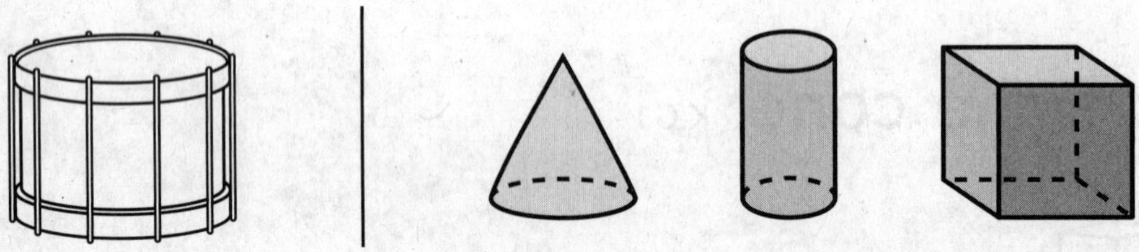

5. Circle the shape that is below the line. K.G.1

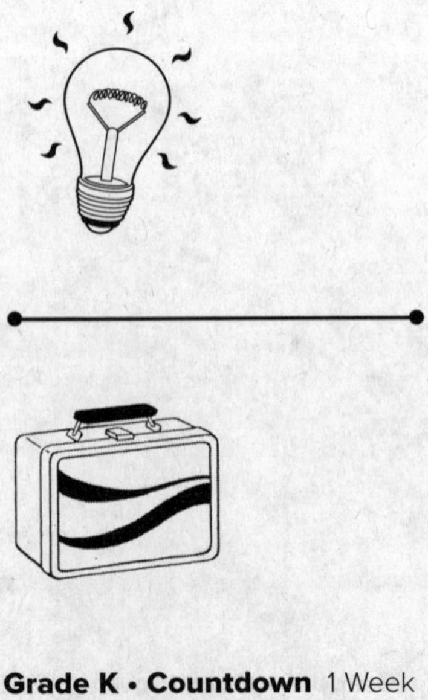

NAME ________________ DATE ________________

Chapter 1 Test

SCORE ________________

1. Count the number of balloons. Circle the number that tells how many. K.CC.3, K.CC.5

1   2   3

2. Circle the bowl with zero peas. K.CC.6

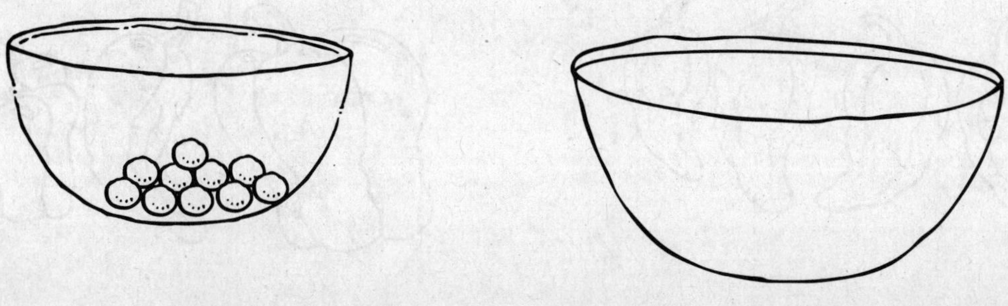

3. Draw three triangles. Write the numeral that tells the number of triangles. K.CC.4a, K.CC.3

4. Circle the group with more peppers. K.CC.6

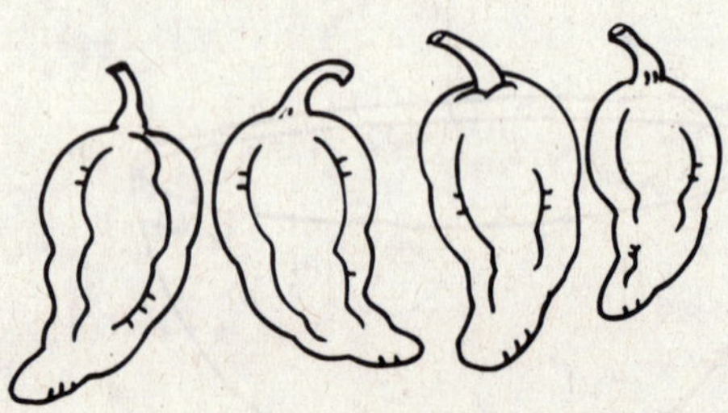

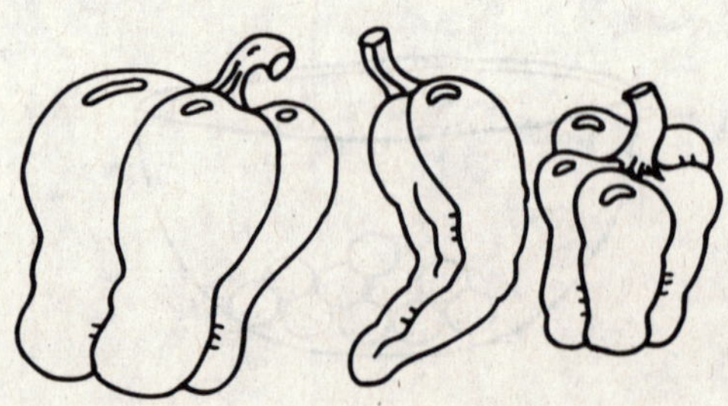

5. Draw a group of flowers that has more than the group shown. K.CC.4a, K.CC.6

THINK SMART FOR SBAC
On the test, you might drag pictures of flowers to an answer box. In this book, you will draw the flowers.

6. Circle the group with five pencils. K.CC.4a, K.CC.1

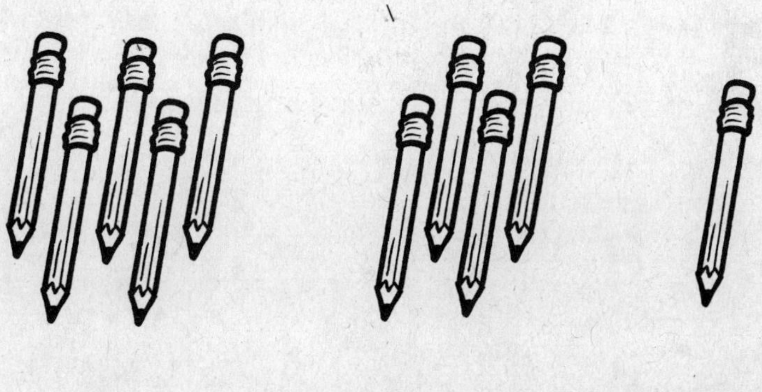

7. Write the number of sandwiches shown. K.CC.3, K.CC.5

8. Write the number of balloons shown. K.CC.3, K.CC.5

9. Draw an X on the groups that show less than 3. K.CC.5, K.CC.6

10. Circle the number that is one more than one. K.CC.4c

0 1 2 3

11. Count the dogs. Find the group of bones with the same number. Draw a line from the group of dogs to the group of bones with the same number. K.CC.4a, K.CC.6

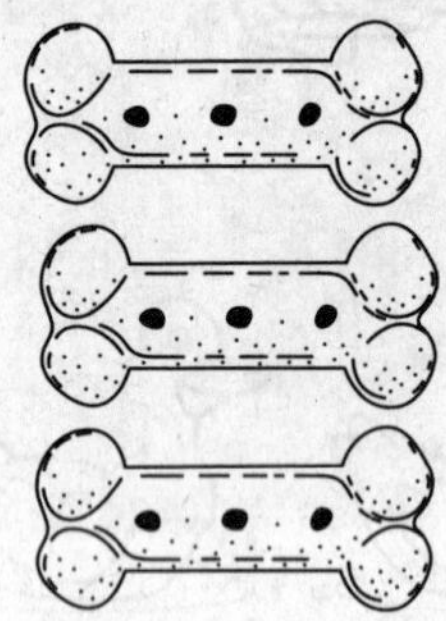

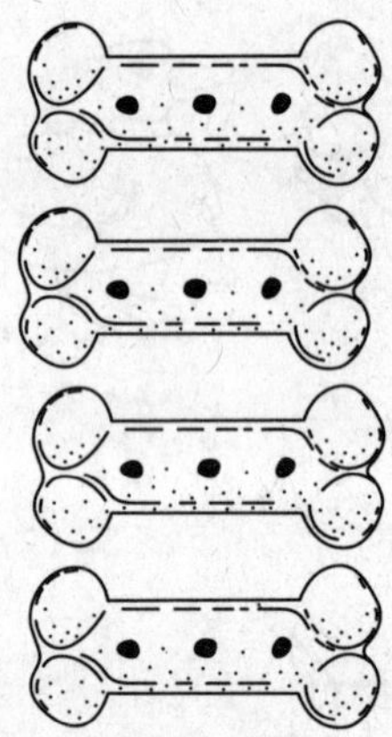

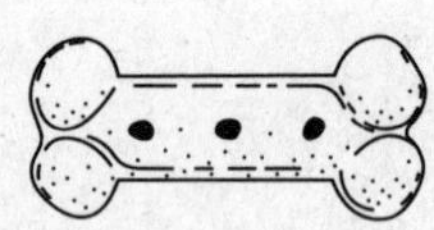

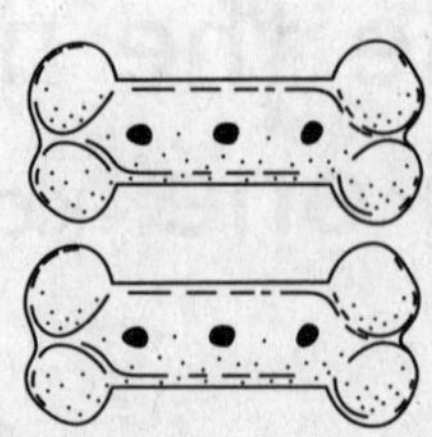

NAME DATE

Chapter 2 Test

SCORE

1. Color the boxes in the ten-frame to show seven. K.CC.4

2. Count the circles. Draw more circles to show eight. Write the number. K.CC.3, K.CC.4

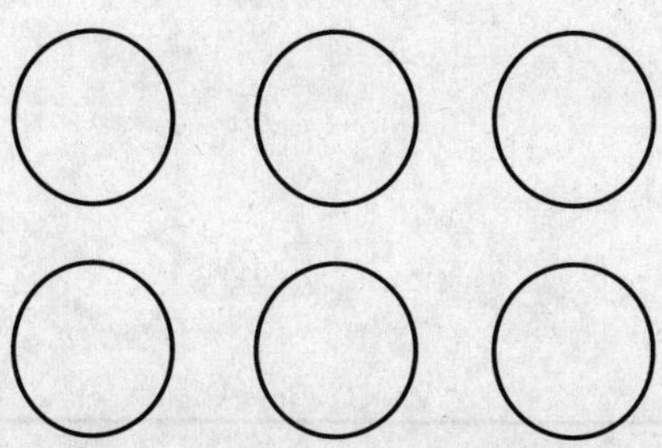

3. Circle the ten-frame that shows nine. K.CC.4b, K.CC.4c

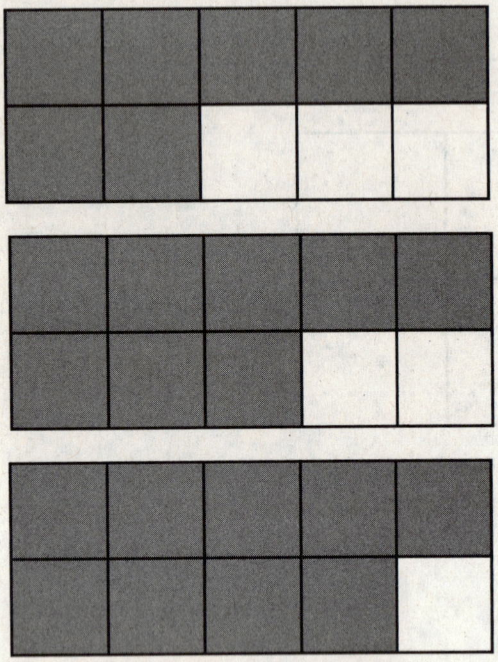

4. Draw ten suns. Write the number. K.CC.3, K.CC.4

5. Count the bunnies. Write the number. Count the carrots. Write the number. Circle the group and number that is greater than the other. K.CC.3, K.CC.6

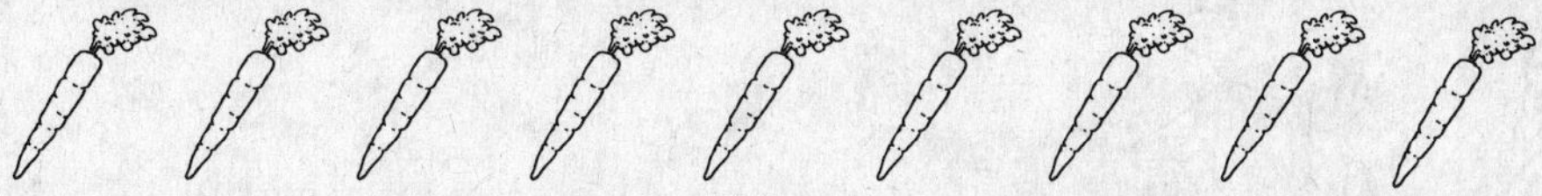

6. Show five and one more pencils on the ten-frame. Write the number of pencils. K.CC.3, K.CC.4

7. Write the number that matches the number of bears shown. K.CC.4a, K.CC.3

8. Circle the number that is one more than six. K.CC.4c

5 6 7 8

9. Circle the ant that is first in the line of ants leaving the picnic. Draw an X on the ant that is third in line. Draw a line under the ant that is fourth in line. Preparation for K.CC.4a

10. Circle the boat that is seventh in line from the lighthouse. Draw an X on the boat that is fifth in line. Draw a line under the boat that is tenth in line. Preparation for K.CC.4a

NAME .... DATE ....

Chapter 3 Test

SCORE ....

1. Count the ducks. Write how many. K.CC.3

____

2. Circle 16 dolls. K.CC.3

3. Draw more circles to make 12. K.CC.3

4. Connect the numbers in order.
Circle the smallest number. K.CC.1

5. How many dinosaurs are there? Circle the correct number. K.CC.3

16 17 18

6. Continue counting by ones. K.CC.1

87 ____ ____ ____

7. Count and draw lines to match objects in one group with objects in the other. Write the numbers. Circle the group that has more than the other. K.CC.3

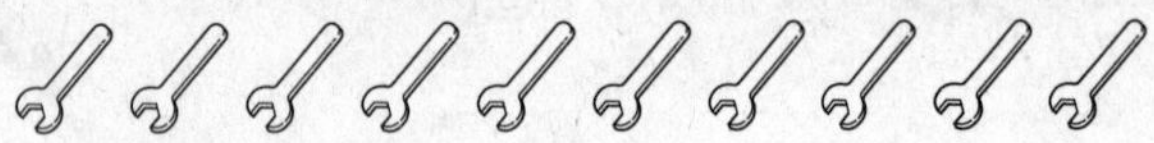

8. Color more boxes in the ten-frame to show 19. K.CC.3

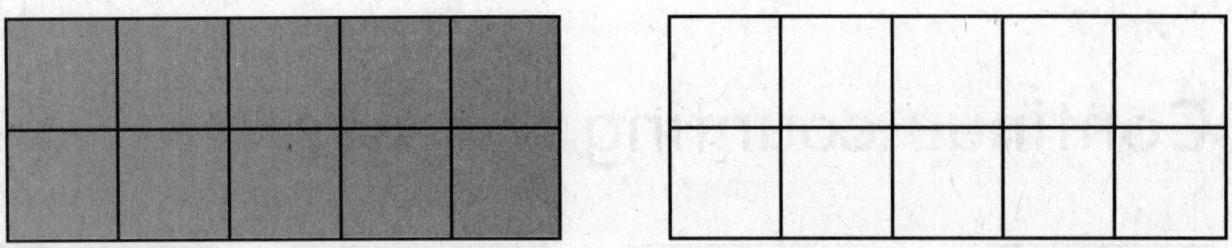

9. Count by tens. What number comes next? Write the number. K.NBT.1

30 40 50 ______

10. Write the number shown by the ten-frames. K.CC.3

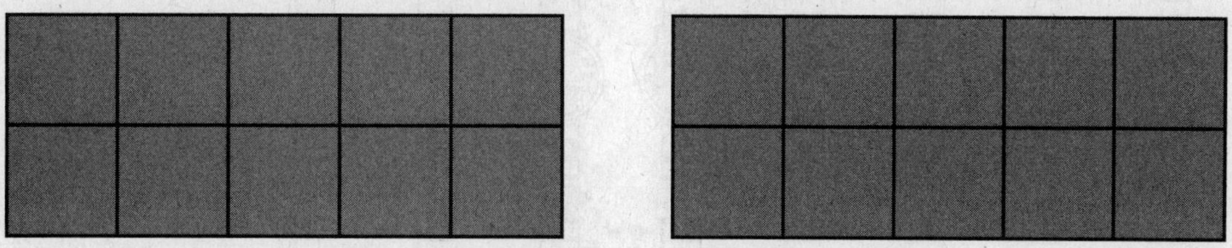

______

11. Circle the number that is one less than 16. K.CC.1

13 14 15

12. Count the watches. Circle the same number of boats. K.CC.1

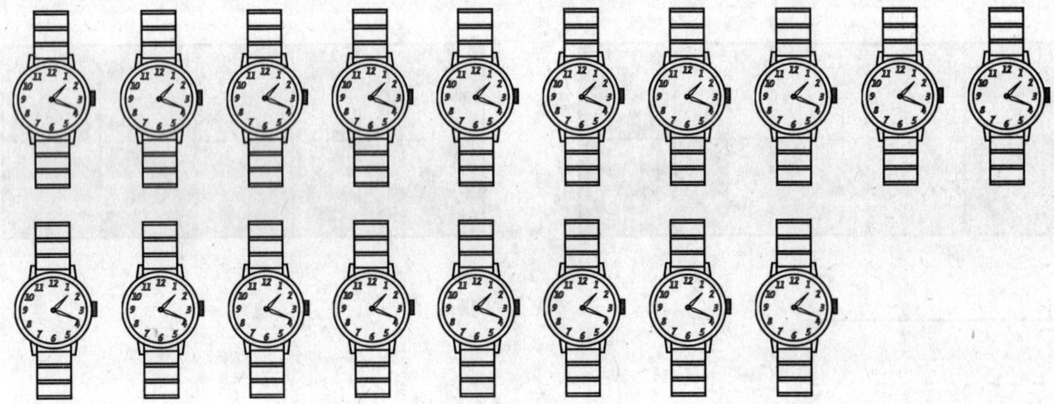

NAME DATE

Chapter 4 Test

SCORE

1. There are 5 umbrellas. Take apart the number. K.OA.3

______ and ______

2. There are 4 flags. Show a way to make 4. K.OA.1

______ and ______

3. Count the ties. Write the number above the ties. K.OA.3

Show a way to take apart the number.

__________ and __________

4. Show a way to make 6 eggs. K.OA.1

__________ and __________

5. There are 8 squares. Break apart a group of 5. Write the number in the other group. K.OA.3

8

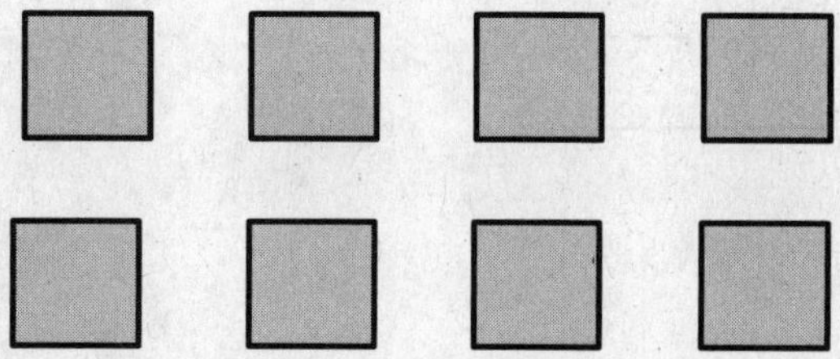

5 and ______

6. Show a way to make 9 by circling screws. Write the numbers. K.OA.1

______ and ______

7. Take apart 10. Write the numbers that make 10. K.OA.3

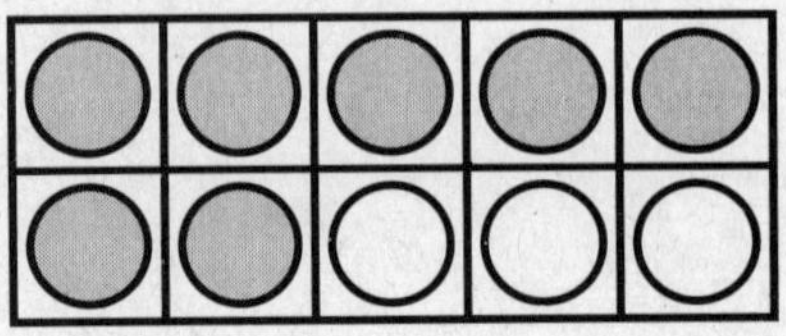

______ and ______

8. Write the number that makes ten. K.OA.4

1 and ______

9. There are 6 shells. Circle shells to make 3 shells and 3 shells. K.OA.1

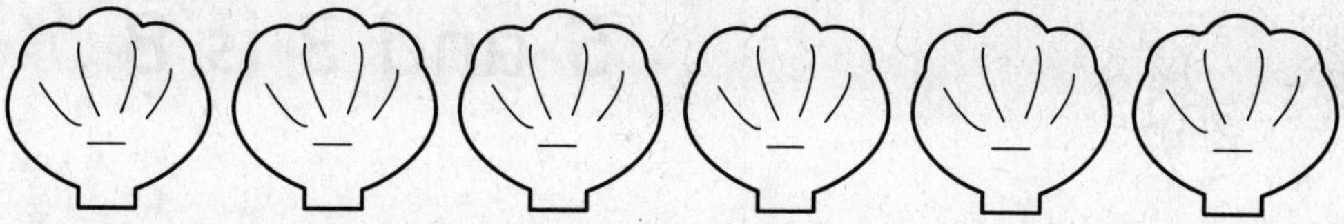

10. Count the faces. Write the number above the faces. K.OA.3

Show how to take apart the number.

5

11. Match the picture to the statement. K.OA.3

5 and 3 is 8.

3 and 3 is 6.

12. Write the number that makes 9. K.OA.1

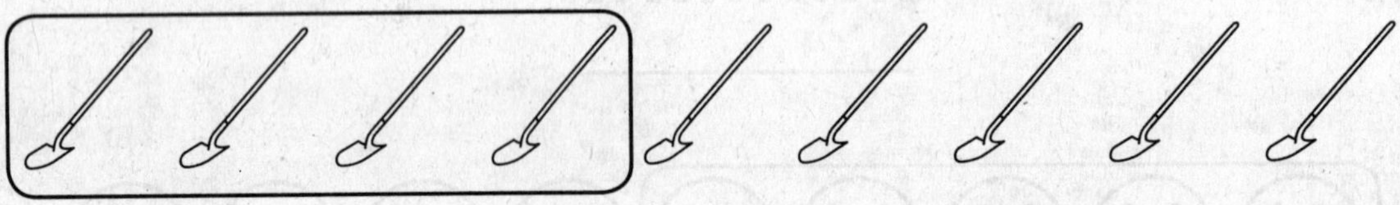

4 and ______

NAME ____________ DATE ____________

Chapter 5 Test

SCORE ____________

1. There are three lizards on the grass. Two more lizards come. How many lizards are there in all? K.OA.1

______

2. Write the number that tells how many there are in all? K.OA.1

4 + 2 = ______ in all

3. Count the number in each group. Write how many there are in all. K.OA.1

________ + ________ = ________

in all

4. Circle two groups to match the number sentence. K.OA.1

$6 + 3 = 9$

5. Count the number in each group. Write the numbers. Then write the number in all. K.OA.1

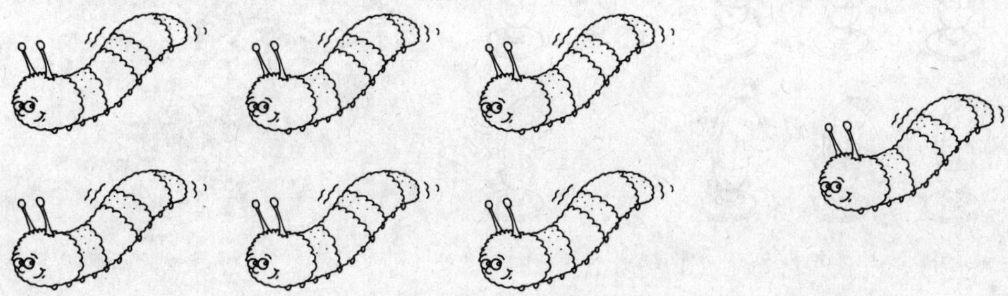

________ + ________ = ________

6. There are 2 cherries. Mrs. Grady buys 2 more. How many cherries are there in all? Write a number sentence. K.OA.1

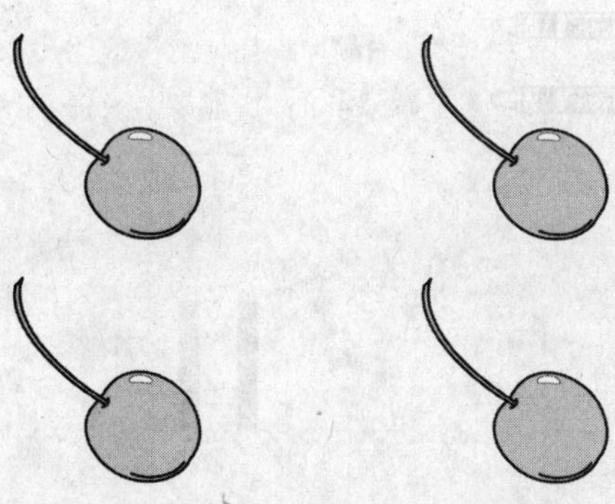

________ + ________ = ________

7. Count the objects and write the numbers. K.OA.4

____ + ____ = ____

8. Huan has 3 crayons. He gets 7 more. How many crayons does he have in all? Circle the correct number. K.OA.4

9 10 11

9. Count the number in each group.
Write how many there are in all. K.OA.1

________ + ________ = ________

in all

10. Circle the plus sign. K.OA.1

+ − × =

11. Circle the picture that shows 6 and 4 make 10. K.OA.4

NAME ______ DATE ______

Chapter 6 Test

SCORE ______

1. Jaylan saw 5 fish. Two swam away. Cross out two fish. Write how many are left. K.OA.1

______

2. There were 9 cookies. Jackson ate 6 cookies. Cross out 6 cookies. Write how many cookies are left. K.OA.1

______

3. Count the number of crabs. Count how many are going away. Count how many are left. Write the numbers. K.OA.1

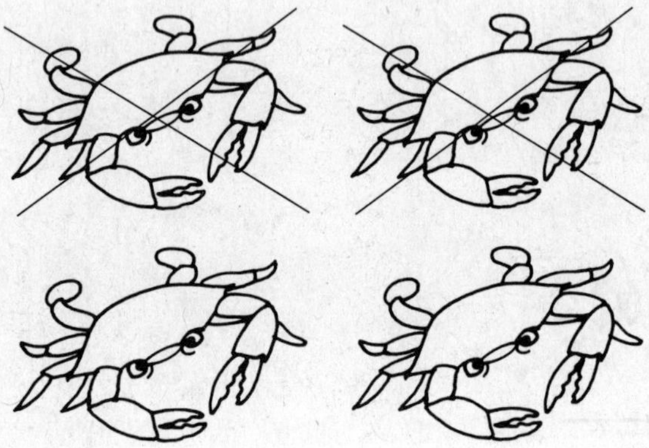

______ – ______ = ______

4. Write a number sentence. K.OA.1

______ – ______ = ______

5. Count the mice. Seven run away. Cross out seven mice. Write how many are left. K.OA.4

______

6. There are 7 jugs of milk. The students drink 2 of them. How many are left? Write a number sentence. K.OA.1

______ − ______ = ______

7. There are ten airplanes. Five airplanes fly away. Cross out five airplanes. Write how many airplanes are left. K.OA.3

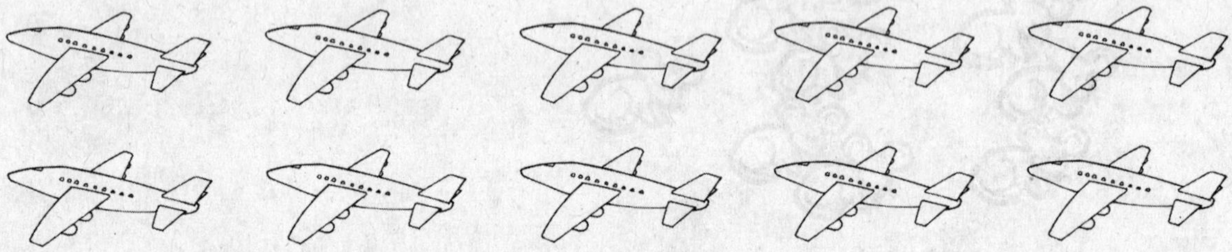

8. Henry had 3 apples. He ate one apple. How many apples does Henry have left? Circle the number. K.OA.1

1 2 3

9. Cross out acorns to show the number sentence. K.OA.1

$$6 - 5 = 1$$

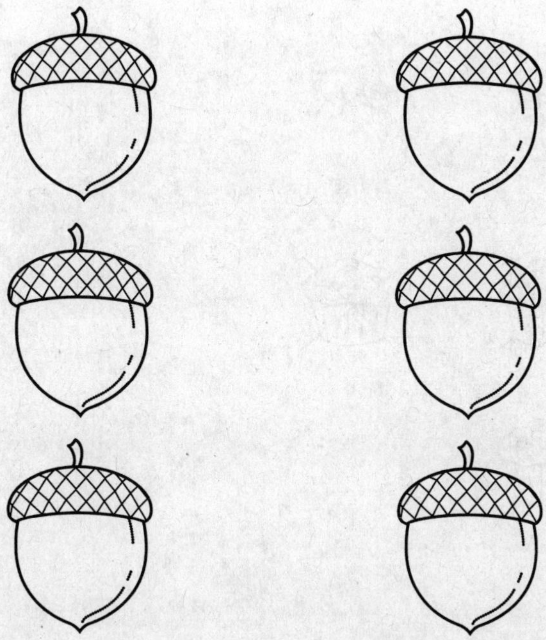

10. Circle the subtraction sign. K.OA.1

$+ \quad - \quad = \quad \times$

11. Circle the picture that shows
8 − 7 = 1. K.OA.4

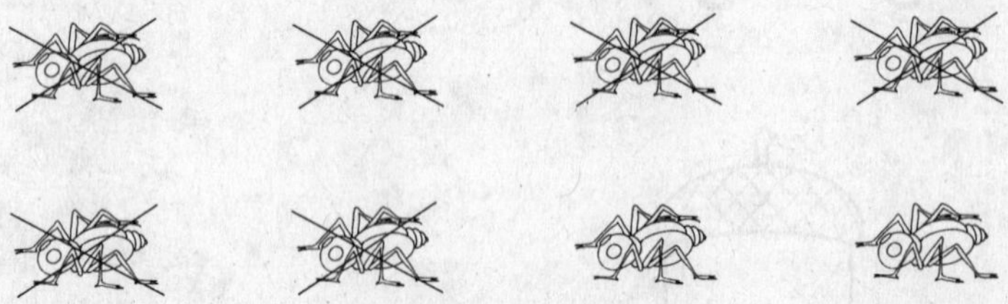

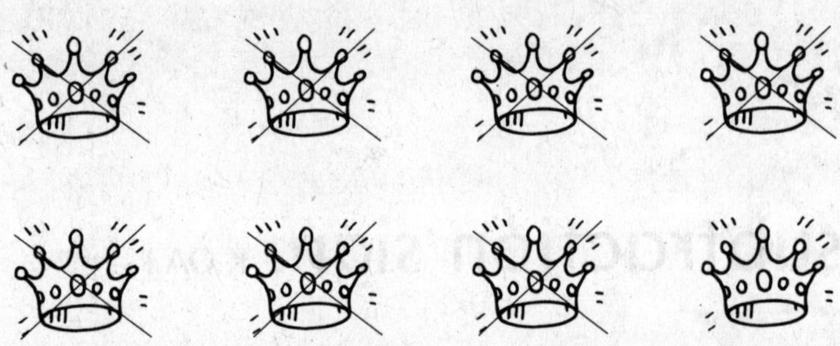

NAME ________________________ DATE ____________

Chapter 7 Test

SCORE ____________

1. Count the number in each ten-frame. Write the numbers. Write how many in all. K.NBT.1

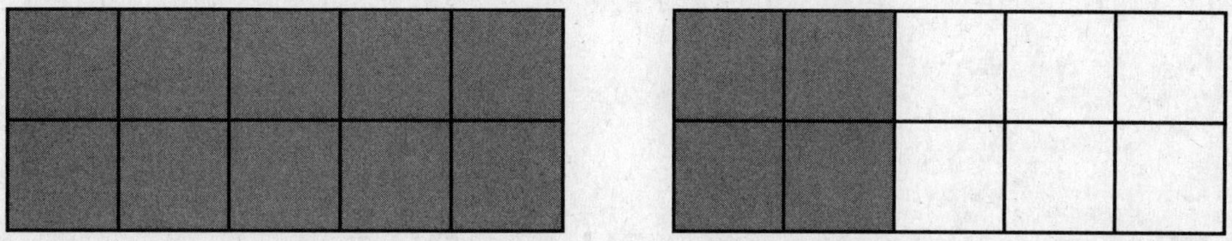

______ and ______ is ______

2. Count the number of each clock. Write each number. K.NBT.1

______	______

3. Shade the ten-frames to show "10 and 7 is 17." K.NBT.1

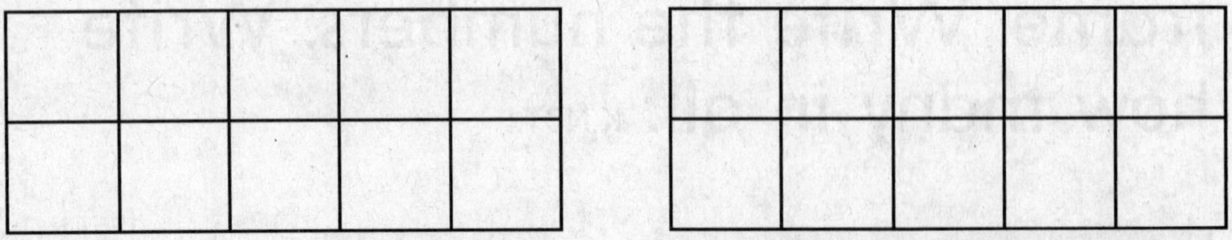

4. There are ten turtles. Six more turtles come. How many turtles are there in all? K.NBT.1

________ and ________ is ________

5. Circle the answer to "ten and five more." K.NBT.1

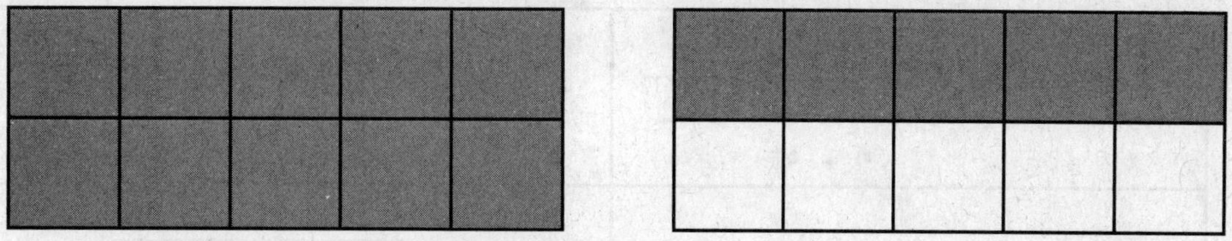

13 14 15

6. There were twelve apples in a tree. 2 apples fell off. Cross out two apples. How many apples are left? K.NBT.1

7. Count the number of each animal. Write each number. K.NBT.1

______	______

8. There are ten horses. Seven more horses come. How many horses are there in all? K.NBT.1

______ and ______ is ______

9. Which is correct? K.NBT.1

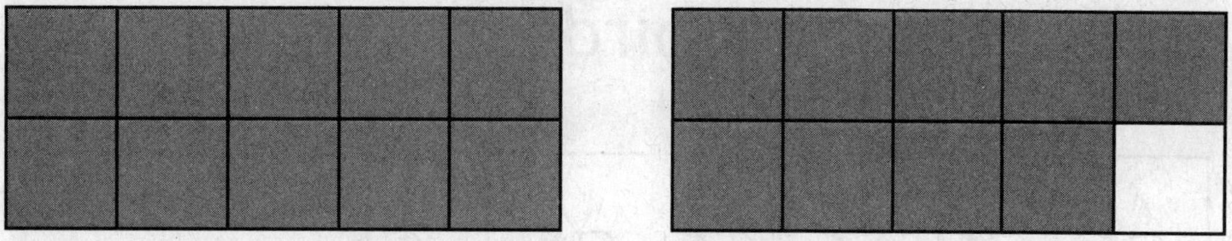

10 and 9 more

10 and 8 more

10 and 7 more

10. There are ten hot air balloons. Ten fly away. Cross out ten balloons. How many are left? K.NBT.1

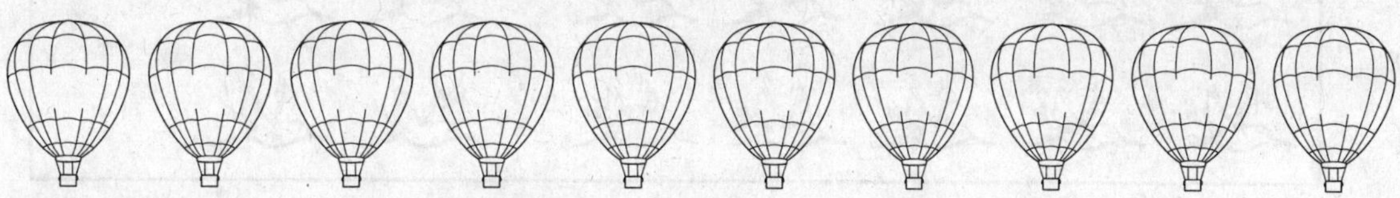

11. Circle the picture that shows

10 birds and 4 birds is 14 birds. K.NBT.1

NAME DATE

Chapter 8 Test

SCORE

1. Compare the objects. Circle the one that is shorter. Underline the one that is longer. K.MD.2

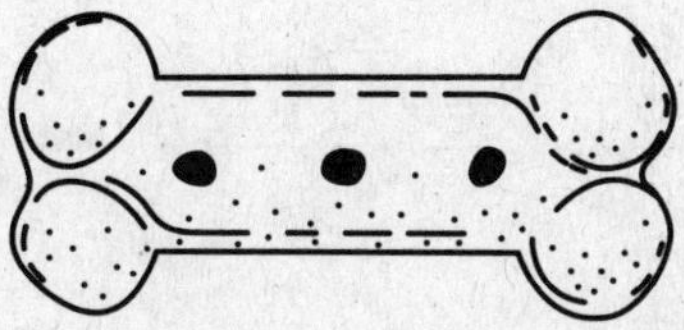

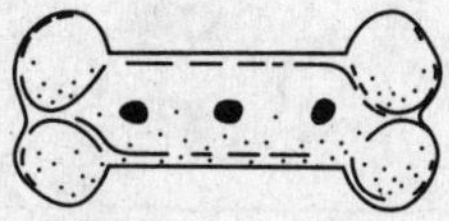

2. Compare the dogs. Circle the taller one. Underline the shorter one. K.MD.2

3. Circle the things that are too heavy to lift. K.MD.2

4. Look at the two objects. Circle the object that is shorter and weighs less than the chair. K.MD.2

5. Match the object to the word. K.MD.2

shorter

longer

6. Circle the bowl that holds more. K.MD.2

7. Circle the lighter object. K.MD.2

8. How many pennies long is the wagon? K.MD.2

____

9. Underline the containers that hold the same amount. K.MD.2

10. Circle the worm that is 6 blocks long. K.MD.2

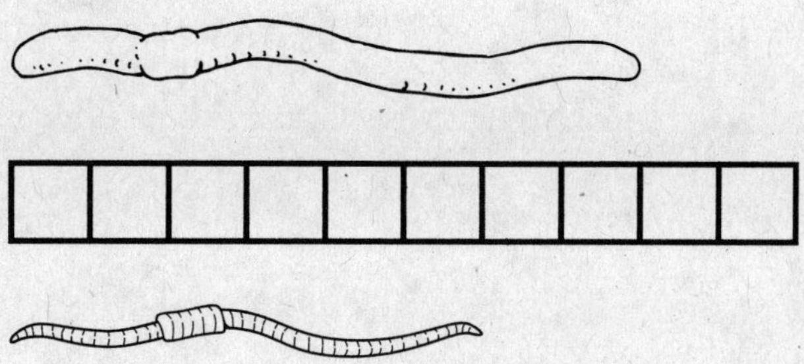

11. Underline the pieces of chalk that are the same length. K.MD.2

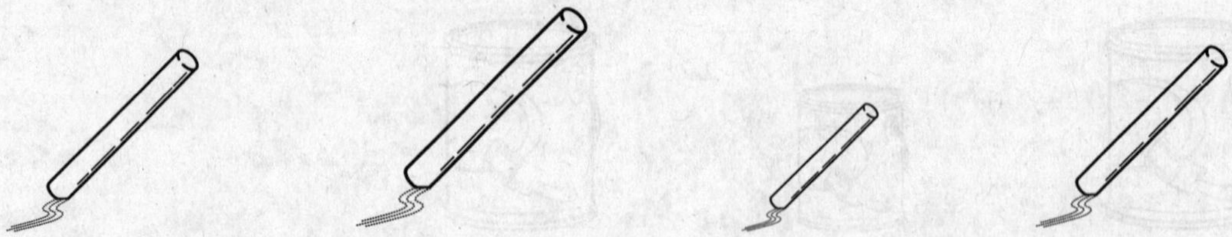

12. How many blocks tall is the cage? K.MD.2

NAME DATE

Chapter 9 Test

SCORE

1. Circle the ones that are alike. K.MD.3

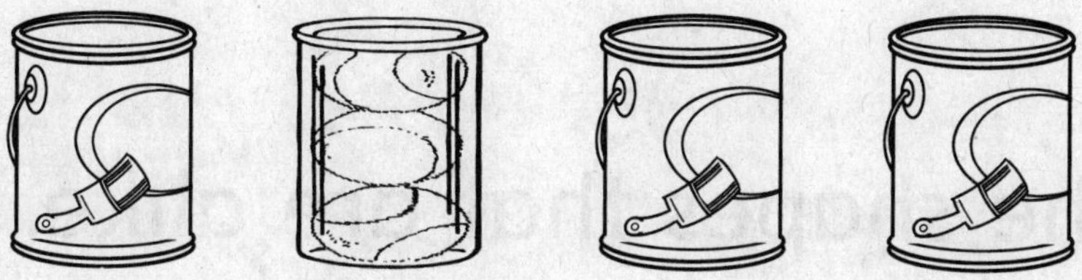

2. Look at the pictures. Circle the pictures that you can eat. K.MD.3

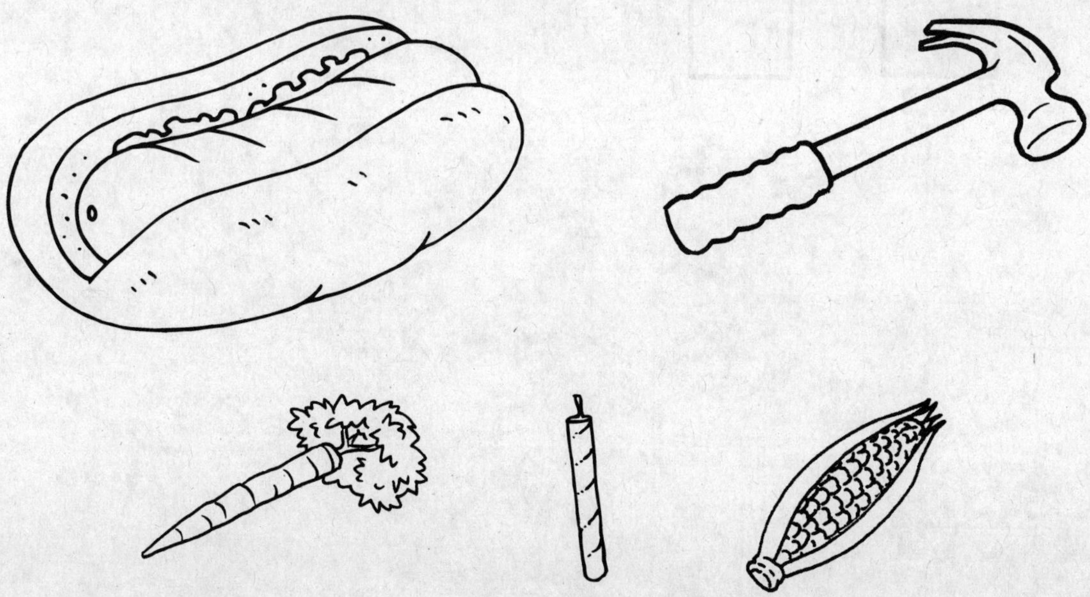

3. Underline the smaller hearts. K.MD.3

4. Circle all the shapes that are alike. K.MD.3

5. Circle the one that is different. K.MD.3

6. Cross out the objects that are not cats. K.MD.3

7. Sort by size. Circle the taller ones. K.MD.3

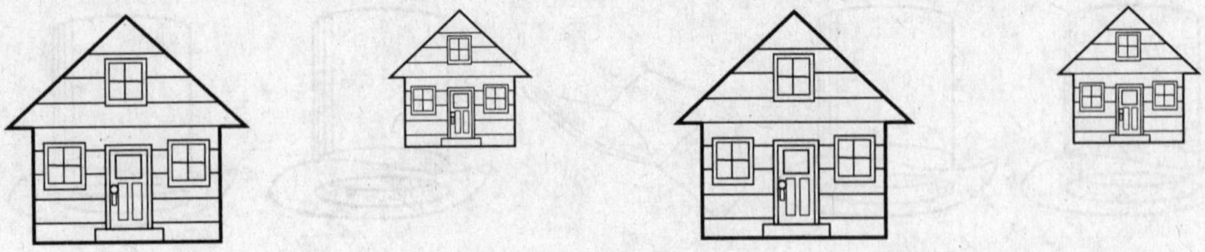

8. Sort by shape. Cross out the one that is not like the others. K.MD.3

9. Sort by number. Circle the groups that have the same number. K.MD.3

10. Sort by color. Underline all the shapes that have the same color. K.MD.3

11. Sort by number. Cross out the group that does not have the same number. K.MD.3

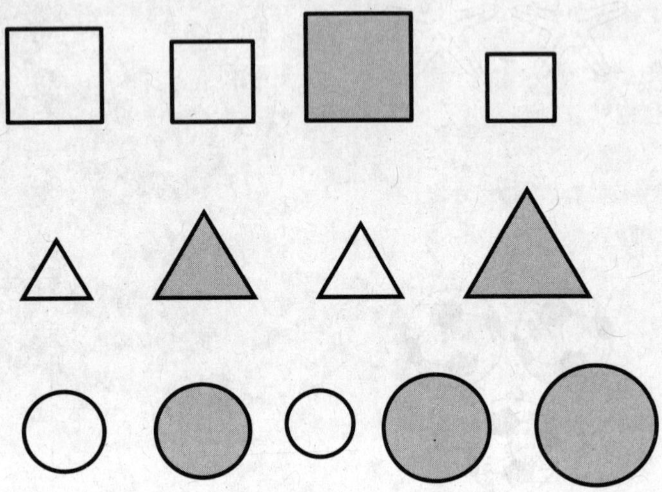

12. Sort by color. Cross out the shape that does not have the same color as the others. K.MD.3

NAME DATE

Chapter 10 Test

SCORE

1. Circle the object that is above the table. K.G.1

2. Circle the dog that is in front of the tree. K.G.1

3. Circle the bug that is nearest to the bucket. K.G.1

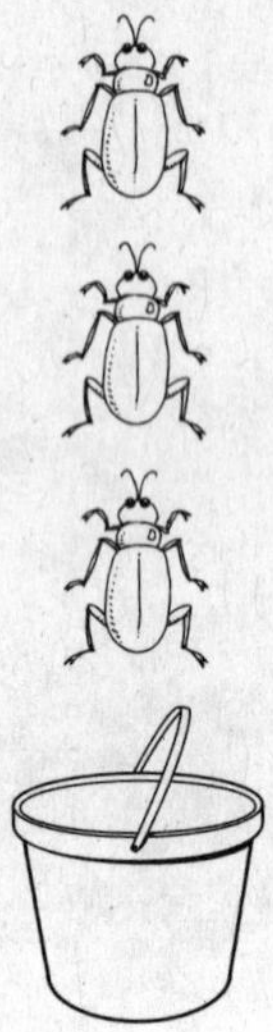

4. Cross out the objects that are below the couch. K.G.1

5. Cross out the objects that are behind the swing set. K.G.1

6. Underline the building that is next to the tallest building. K.G.1

7. Draw a circle above the line. K.G.1

---

8. Are the shoes in front of or behind the door? Circle the answer. K.G.1

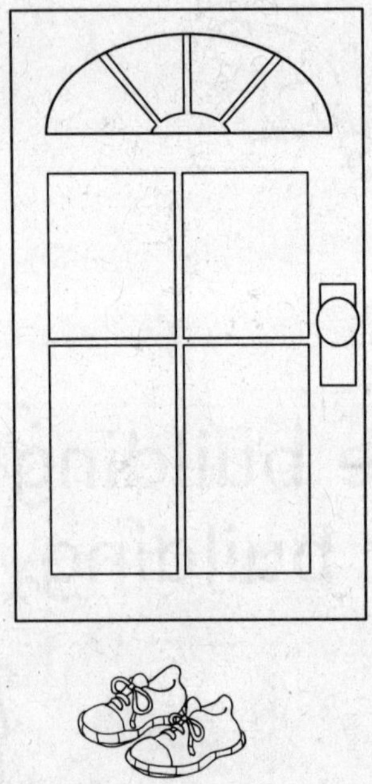

in front of behind

9. Circle any butterfly that is next to the flower. K.G.1

10. Shade the rectangle below the line. K.G.1

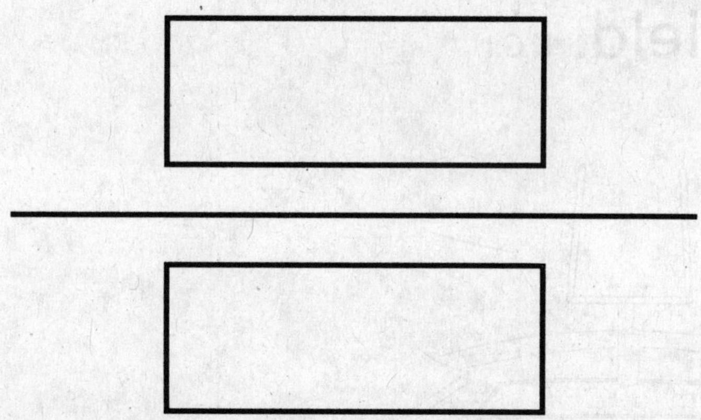

11. Match the two snowmen to the words. K.G.1

in front of

behind

12. Circle the football that is nearest to the field. K.G.1

NAME DATE

Chapter 11 Test

SCORE

1. Shade the squares. K.G.2

2. Underline the object that has a hexagon shape. K.G.2

3. Draw the shape that comes next. K.G.4

4. Circle the shapes that are used to make the design. K.G.5

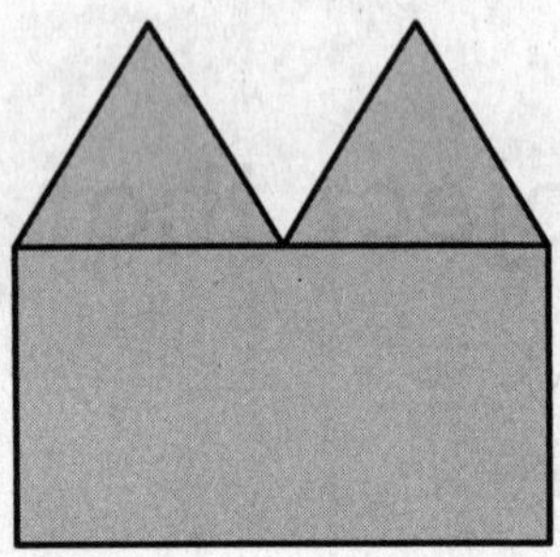

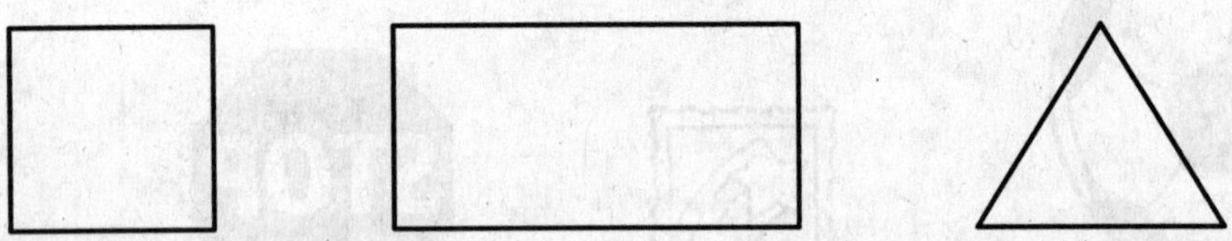

5. Cross out the pictures that do not have circles. K.G.2

6. Underline the shape that is most like a rectangle. K.G.2

7. Count the number of circles. K.G.4

8. Circle the name of the shape that is next to the circle. K.G.1

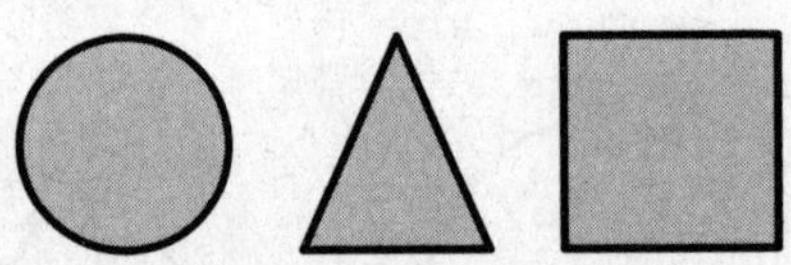

triangle hexagon square

9. Draw the shape that comes next. K.G.4

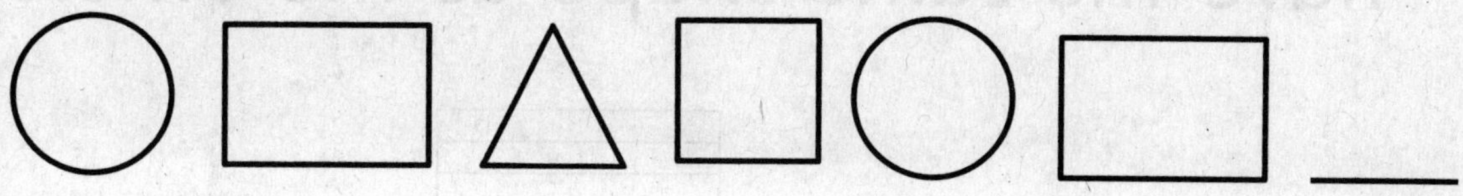

10. Cross out the shapes that are not used to make the design. K.G.5

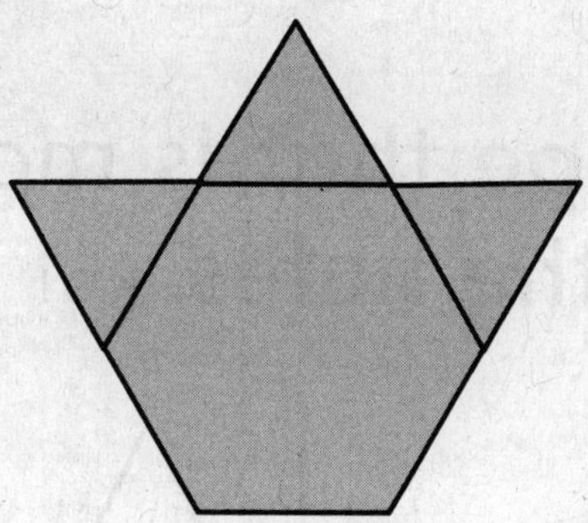

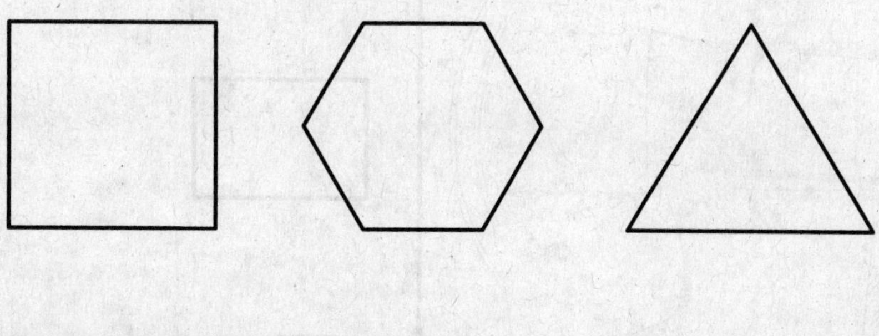

11. Cross out the object that does not have the same shape as the others. K.G.1

12. Circle the shape that is most like the shape of the notes. K.G.1

NAME DATE

Chapter 12 Test

SCORE

1. Put an X on the shapes that are not spheres. K.G.2

2. Circle the shape that is most like a cone. K.G.1

3. Underline the shapes that can slide. K.G.4

4. Circle the shape that best matches the tomato. K.G.2

5. Circle the three-dimensional shapes. K.G.1, K.G.2

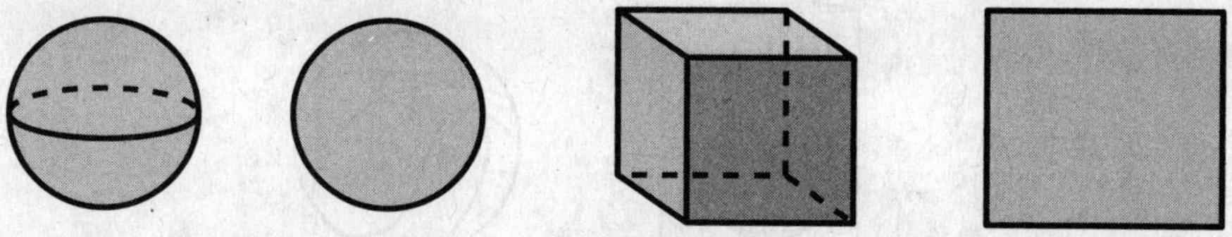

6. Underline the three-dimensional shape that is made from all squares. K.G.2

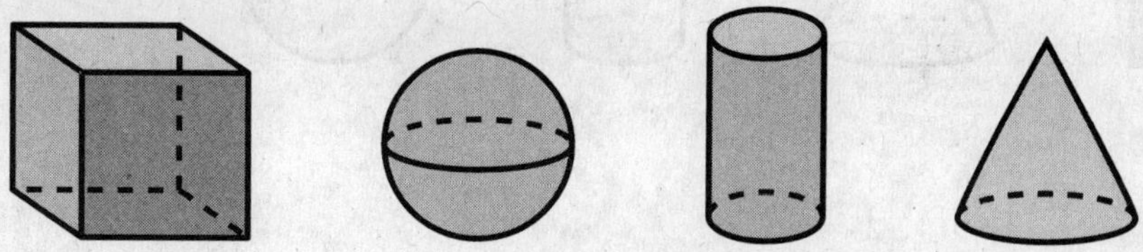

7. Cross out the shape that will not roll. K.G.4

8. Circle the shapes can be stacked. K.G.4

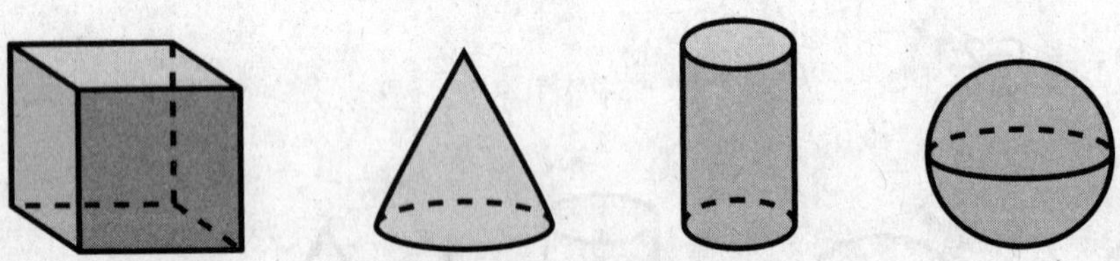

9. Cross out the shape that is different. K.G.2

10. Underline the shapes that make the can and the tennis balls. K.G.2

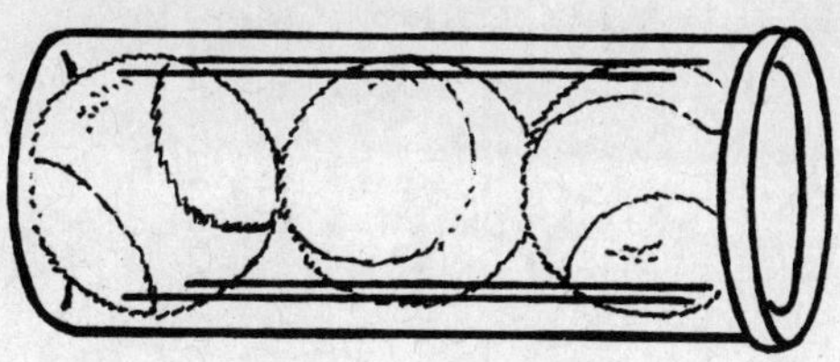

11. Count the spheres in the basket. Write the number. K.G.2

12. Cross out the shapes that are not in the picture. K.G.2

NAME ______ DATE ______

SCORE ______

Performance Task

Making Dinner

Mrs. Jackson is making dinner.

Write your answers on another piece of paper. Show all your work to receive full credit.

Part A

K.CC.6

Circle the bowl that has zero cherries.

Part B

K.CC.3, K.CC.5

Count the number of broccoli pieces. Write the number.

______

Performance Task *(continued)*

K.CC.3, K.CC.5

Part C

Count the number of forks. Circle the number that is one more than that.

3 4 5

K.CC.1, K.CC.4

Part D

Circle the group with 5 tomatoes.

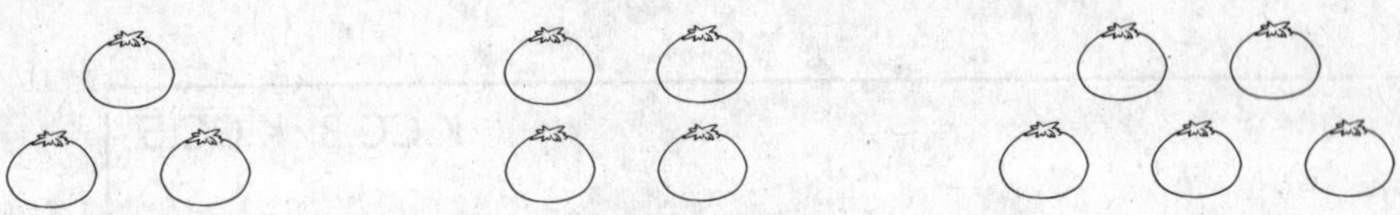

K.CC.1, K.CC.4

Part E

Circle the group with more cupcakes.

NAME DATE

SCORE

Performance Task

At the Cat Shelter

You are helping at a cat shelter.

Part A

These are the cats that need help.
Write the number of cats.

______

Part B

You have these bowls for cat food.
Write the number of bowls.

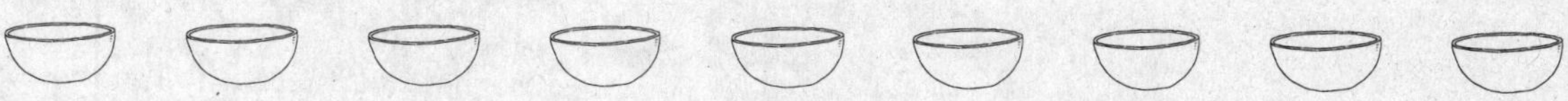

______

Performance Task *(continued)*

Part C

Match the cats to the bowls.

Part D

Circle the group that is less than the other group.

Part E

Draw one ball for each cat.

NAME DATE

SCORE

Performance Task

At the Hat Shop

You are putting hats in boxes at a hat shop.

Write your answers on another piece of paper. Show all your work to receive full credit.

Part A

K.CC.3, K.CC.5

These are the hats that need boxed. Write the number of hats.

Part B

K.CC.3, K.CC.5

These are the boxes. Write the number of boxes.

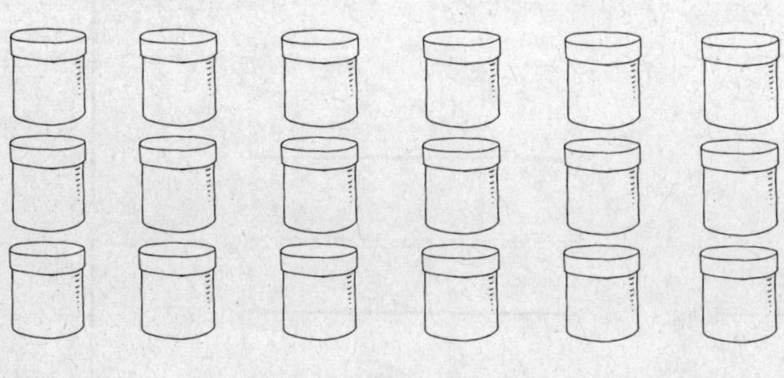

Performance Task *(continued)*

K.CC.3, K.CC.6

Part C

Match the hats to the boxes. Circle the group that has more.

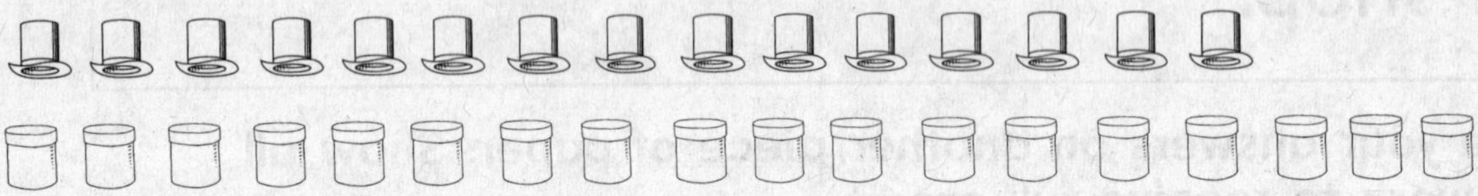

K.CC.3, K.CC.5

Part D

There are 20 price tags. Color the ten-frames to show the number of tags.

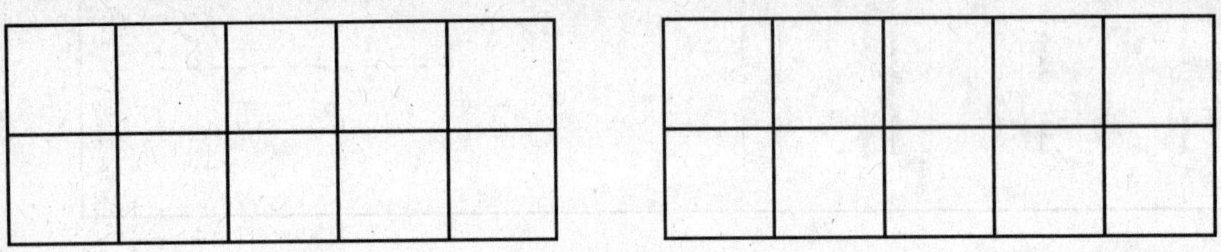

K.CC.3, K.CC.5

Part E

Count the price tags by 10.

10

NAME

DATE

SCORE

Performance Task

Raising Rabbits

You are raising rabbits.

Write your answers on another piece of paper. Show all your work to receive full credit.

Part A

K.OA.1

There are six rabbits. Show two different ways to make 6.

________ and ________

________ and ________

Part B

K.OA.1

Show a way to make 8 cabbages. Write the numbers.

________ and ________

Performance Task *(continued)*

K.OA.3

Part C

There are 7 carrots. Break apart the carrots into two groups. Circle your groups. Write the numbers.

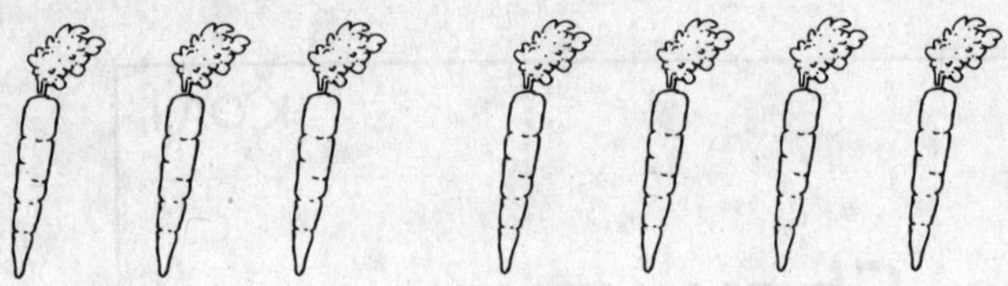

_____ and _____

K.OA.4

Part D

There are ten ears of corn. Write the other number that makes ten.

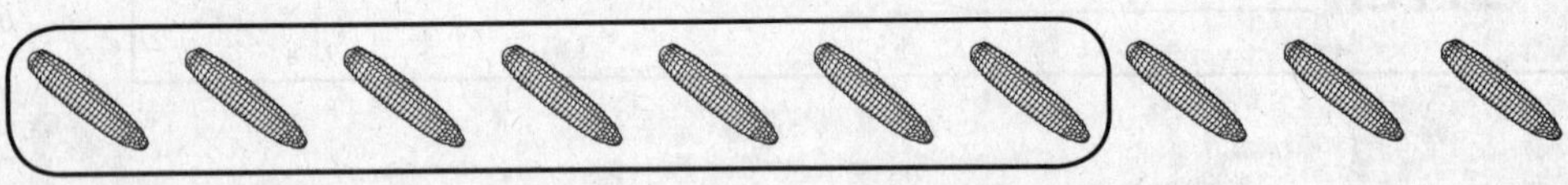

7 and _____

NAME ______________________ DATE ______________

SCORE ______________

Performance Task

At the Sports Store

You are working at a sports store. The store sells different kinds of balls.

Write your answers on another piece of paper. Show all your work to receive full credit.

Part A

K.OA.1, K.CC.3

There are 5 soccer balls on the floor. There are 3 more on the shelf. How many are there in all?

______ + ______ = ______

Part B

K.OA.1, K.CC.3

Circle two groups of baseballs to show 8 and 1 more is 9.

Performance Task *(continued)*

K.OA.1, K.CC.3

Part C

Count the number of footballs in each group. Write the numbers. Write how many in all.

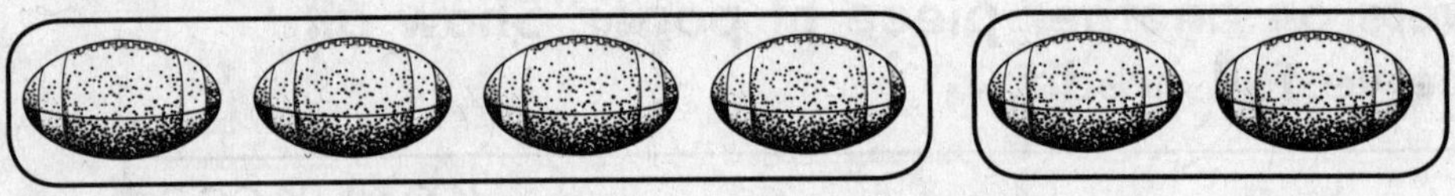

______ + ______ = ______

K.OA.1, K.CC.3

Part D

You sell 12 tennis balls on Monday. You sell 0 tennis balls on Tuesday. How many tennis balls are sold in all? Write the number.

______ + ______ = ______ in all

NAME ______________________ DATE ______________

SCORE ______________

Performance Task

At the Park

You count animals while you are at the park.

Write your answers on another piece of paper. Show all your work to receive full credit.

Part A

K.OA.I, K.CC.3

There are six squirrels. 3 run away. Cross out 3 squirrels. How many are left?

____ − ____ = ____

Part B

K.OA.I, K.CC.3

There are 4 cats. One cat leaves. Circle the number that shows how many cats are left.

2 3 4

Performance Task *(continued)*

K.OA.I, K.CC.3

Part C

There are 9 birds. Some fly away. Write a number sentence to find how many are left.

______ − ______ = ______

birds left

K.OA.I, K.CC.3

Part D

The squirrels take some acorns. Circle the number sentence that shows the acorns the squirrels took.

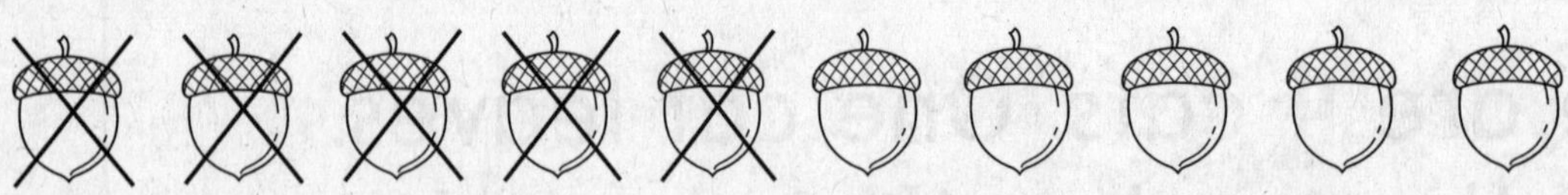

$10 - 5 = 5$ $9 - 5 = 5$ $10 - 6 = 4$

NAME ______________________ DATE ______________

SCORE ______________

Performance Task

At the Airport

At the airport, you are watching planes take off.

Write your answers on another piece of paper. Show all your work to receive full credit.

Part A

K.NBT.1, K.OA.1

Count the numbers in each ten-frame. Write the numbers. Write how many in all.

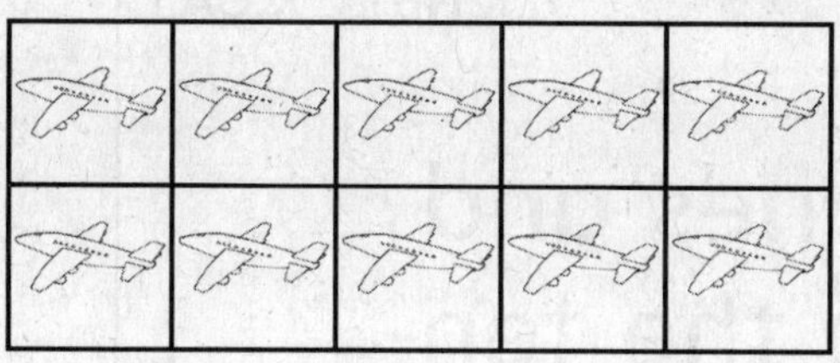

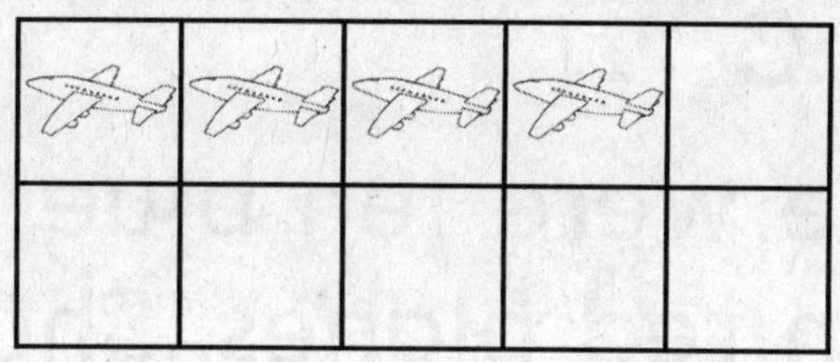

______ + ______ = ______

Performance Task *(continued)*

K.NBT.1, K.OA.1

Part B

There were twelve airplanes. Two airplanes left. Cross out two airplanes. How many airplanes are left?

________ airplanes left

K.NBT.1, K.OA.1

Part C

There were ten blue planes and seven red planes. Shade the ten-frames to show how many planes in all. Write the number.

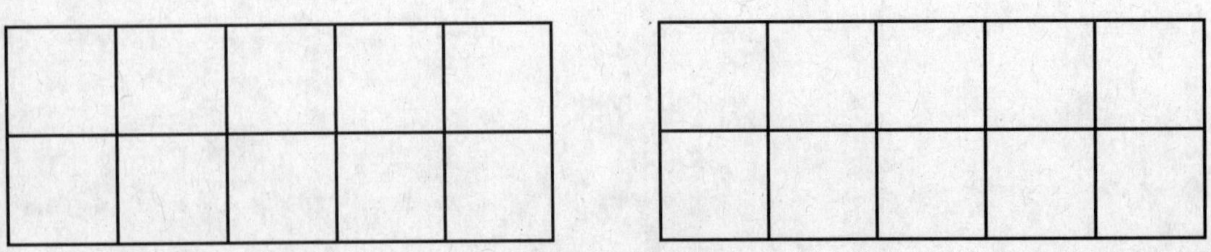

____

NAME DATE

SCORE

Performance Task

Baking Bread

You are baking bread with your mother.

Write your answers on another piece of paper. Show all your work to receive full credit.

Part A

K.MD.2

Circle the mixing bowl that holds more. Cross out the one that holds less.

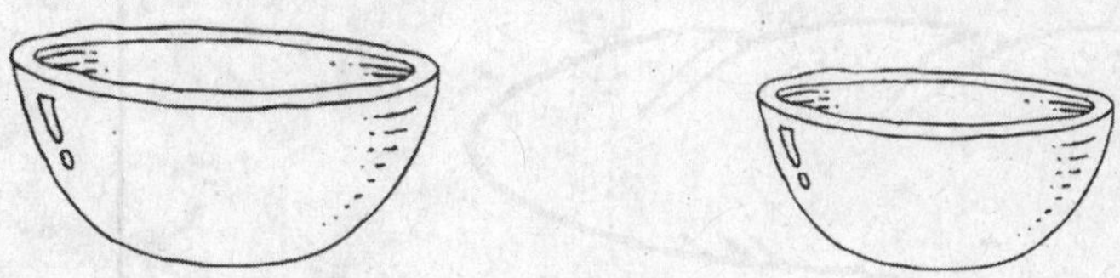

Part B

K.MD.2

Circle the loaf of bread that is longer than the tray.

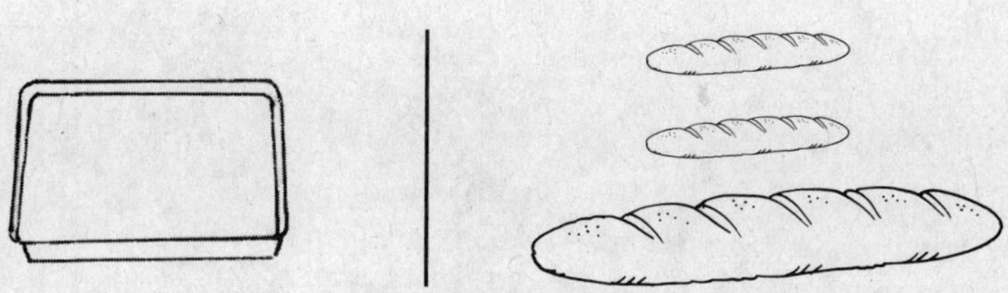

Performance Task *(continued)*

K.MD.2

Part C

Cross out the kitchen object that is too heavy to lift by yourself.

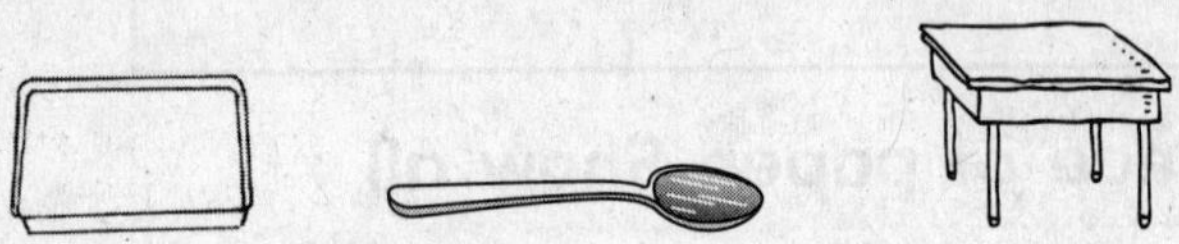

K.MD.2, K.CC.3, K.CC.4

Part D

How many pennies long is the loaf of bread? Write the number.

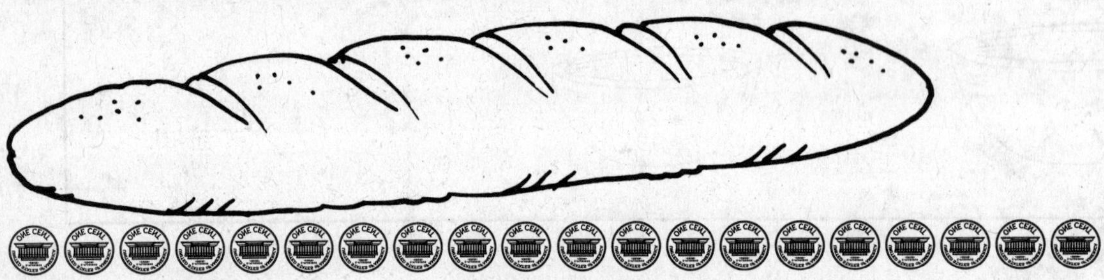

____

NAME .... DATE ....

SCORE ....

Performance Task

A Trip to the Grocery Store

You and your father take a trip to the grocery store. Research how many objects are in a dozen.

Write your answers on another piece of paper. Show all your work to receive full credit.

Part A

K.MD.3

Circle the carrots that are alike.

Part B

K.MD.3

Cross out the items that you cannot eat.

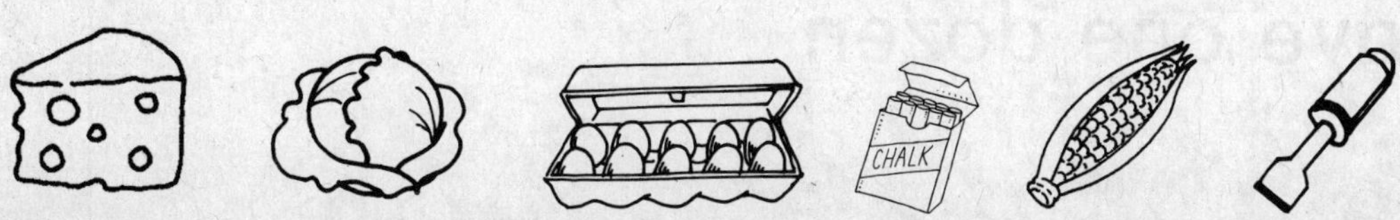

Performance Task *(continued)*

K.MD.3

Part C

Sort by size. Underline the larger muffins.

K.MD.3

Part D

Sort by shape. Circle all the ones that are alike.

K.MD.3, K.CC.3

Part E

Sort by number. Circle the objects that have one dozen.

NAME ____________ DATE ____________

SCORE ____________

Performance Task

A Scavenger Hunt

You are looking for some objects for a scavenger hunt.

Write your answers on another piece of paper. Show all your work to receive full credit.

Part A

K.G.1, K.CC.3

How many objects are below the table? Circle them. Write the number.

____

Part B

K.G.1, K.CC.3

Underline the feather that is nearest to the duck.

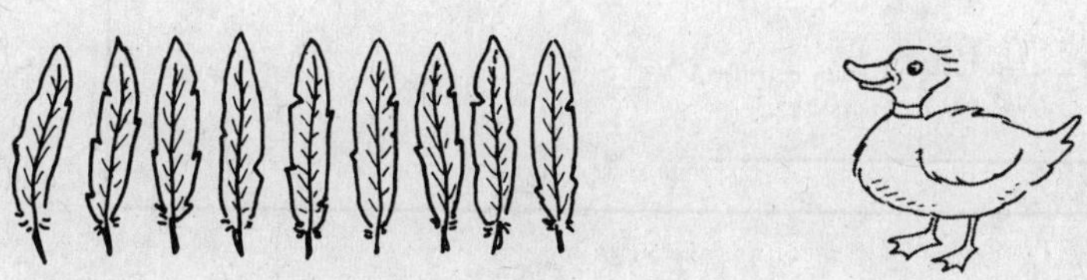

Performance Task *(continued)*

K.G.I, K.CC.3

Part C

How many objects are behind the teepee? Circle them. Write the number.

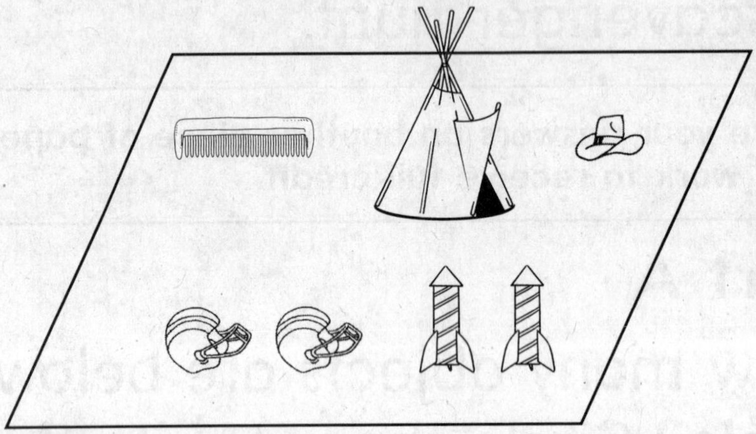

K.G.I, K.CC.3, K.OA.I

Part D

There are some objects above the desk. You take away three of the objects. Cross out three of the objects above the desk. How many are left above the desk? Write the number.

NAME

DATE

SCORE

Performance Task

Building with Tiles

Marcello has tiles that are different shapes.

Write your answers on another piece of paper. Show all your work to receive full credit.

Part A

K.G.2

The different tiles are shown. Circle the triangle. Underline the hexagon. Put an "X" on the square.

Part B

K.G.4

Marcello makes a pattern with his tiles. Draw the two tiles that come next.

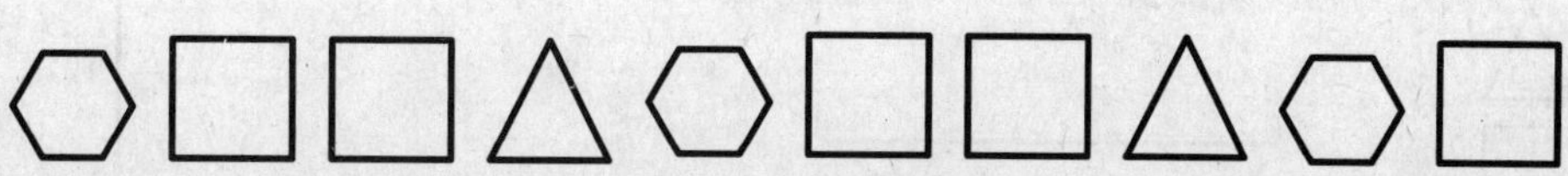

___ ___

Performance Task *(continued)*

K.G.5, K.CC.3

Part C

Marcello builds a star design out of tiles. How many of each tile does he use? Write the numbers.

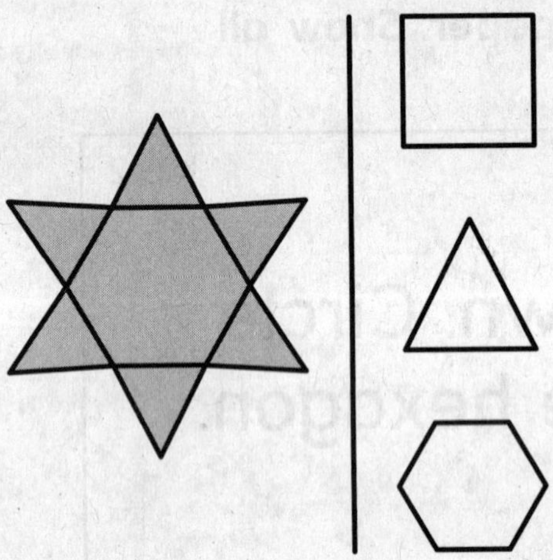

_____

_____

_____

Part D

K.G.2, K.OA.4

Marcello has these shapes. He uses 8 of his triangles in another design. Cross out 8 triangles. How many triangles are left? Write the number.

_____ triangles left

NAME

DATE

SCORE

Performance Task

Cleaning Out the Pantry

There are several objects in Mrs. Hernandez's pantry.

Write your answers on another piece of paper. Show all your work to receive full credit.

Part A

K.G.2

Match each object to the three-dimensional shape.

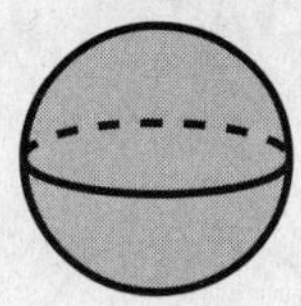

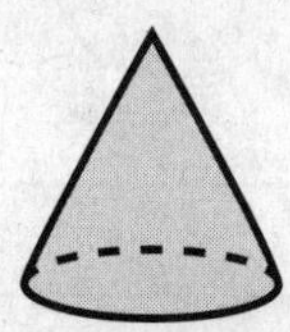

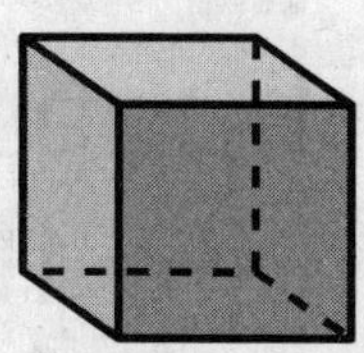

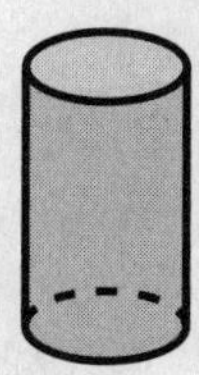

Performance Task *(continued)*

K.G.4, K.CC.3

Part B

Underline the shapes that will roll.
Count them. Write the number.

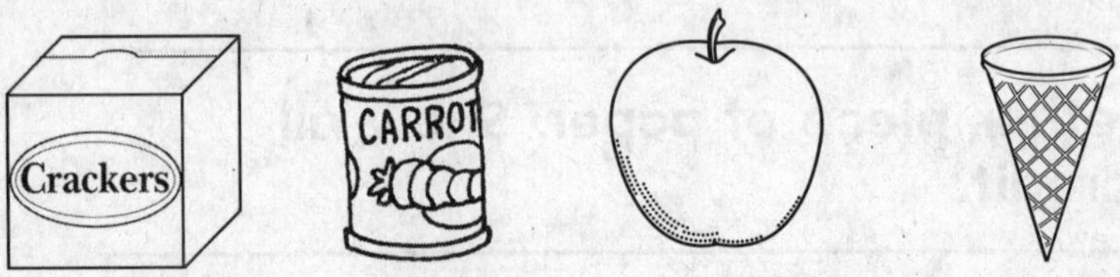

K.G.4, K.CC.3

Part C

Circle the shapes that will stack.
Count them. Write the number.

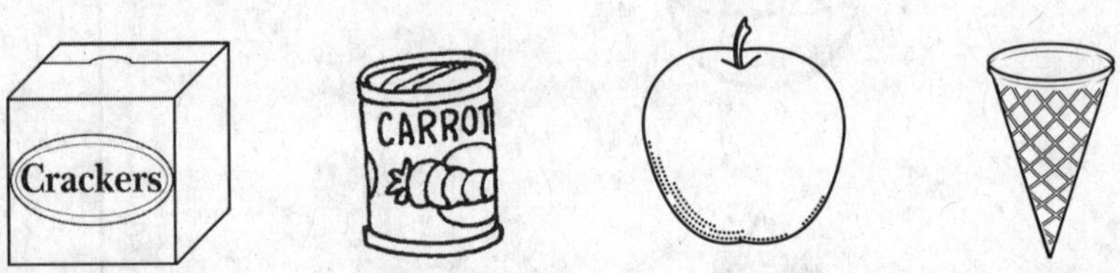

K.G.4, K.CC.3

Part D

Underline the shapes that will slide.
Count them. Write the number.

NAME ______________________ DATE ______________

Benchmark Test 1

SCORE ______________

1. Circle 18 coats. K.CC.3

2. Count the horseshoes. Write the number above the horseshoes. K.OA.3

____

Show a way to take apart the number.

______________ and ______________

3. Circle the number that is one more than 7. K.CC.4c

6 7 8 9

4. Circle the basket with zero apples. K.CC.6

5. Connect the numbers in order. Circle the largest number. K.CC.1

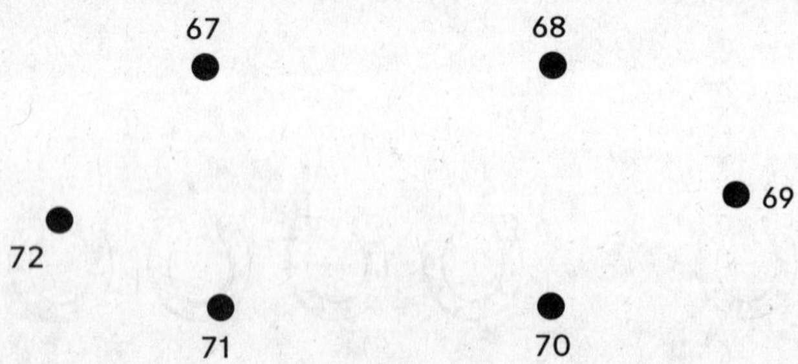

6. Show a way to make 7 by circling houses. Write the numbers. K.OA.1

__________ and __________

7. Count the dots. Draw more dots to show nine. Write the number. K.CC.3, K.CC.4

8. Circle the deer that is first in line to go to the pond. Draw an X on the deer that is third in line. Draw a line under the deer that is last in line. K.CC.4a

9. How many hats are there? Circle the right number. K.CC.3

14 15 16

10. Write the number that makes ten. K.OA.4

3 and ______

11. Write the number that matches the number of horns shown. K.CC.4a, K.CC.3

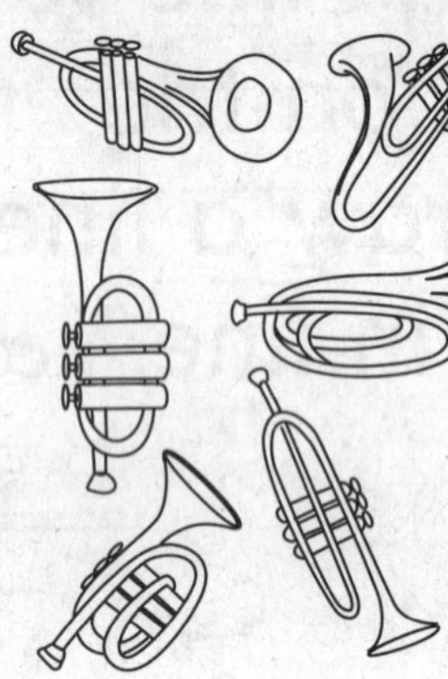

______

12. Circle the group with more bees. K.CC.6

13. Count the wagons. Write the number. Count the wheelbarrows. Write the number. Circle the group and number that is greater than the other. K.CC.3, K.CC.6

14. Color more boxes in the ten-frame to show sixteen. Write the number. K.CC.3

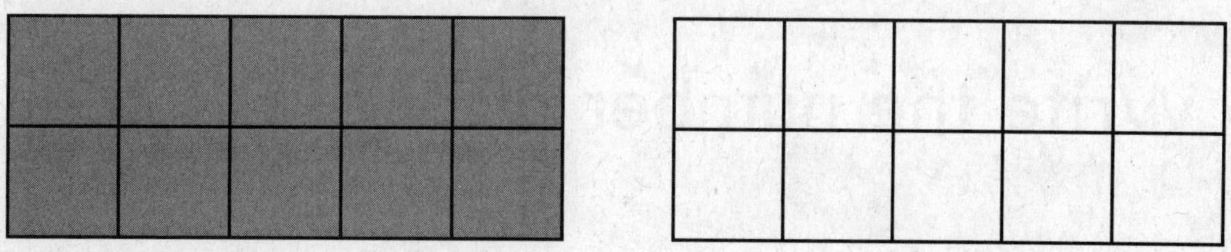

15. Match the picture to the statement. K.OA.3

4 and 1 is 5

4 and 5 is 9

16. Shade the ten frame to make eight. Write the number. K.CC.4

17. Write the number of birds. K.CC.3, K.CC.5

18. Circle the number that is one more than 12. K.CC.1

13 14 15

19. There are 12 squares. Break apart a group of 8. Write the number in the other group. K.OA.3

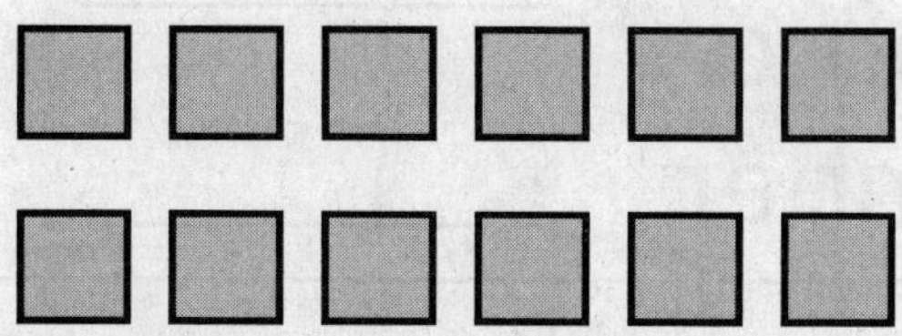

8 and ______

20. Circle the group with two flowers. K.CC.4a, K.CC.1

Performance Task

Building Pianos

Mr. Havill is a piano builder. Research the number of keys on a piano.

Write your answers on another piece of paper. Show all your work to receive full credit.

K.CC.3

Part A

How many keys are on a piano? Write the number.

K.CC.3

Part B

Write the next three numbers.

K.CC.4

Part C

A piano takes 13 days to ship. It takes 1 more day to unpack. What number is 1 more than 13?

NAME ________________ DATE ________________

Benchmark Test 2

SCORE ________________

1. There are two pigs in a pen. Three more pigs come in the pen. How many pigs are there in all? K.OA.1

____

2. Jaylan saw 6 bugs. Two crawled away. Cross out two bugs. Write how many are left. K.OA.1

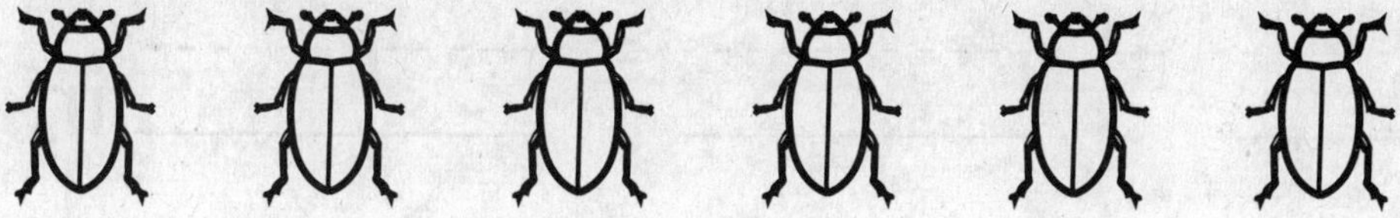

____

3. Circle the number that shows "10 and 2 more." K.NBT.1

2 10 11 12

4. Count the numbers in each ten frame. Write the numbers. Write how many in all. K.NBT.1

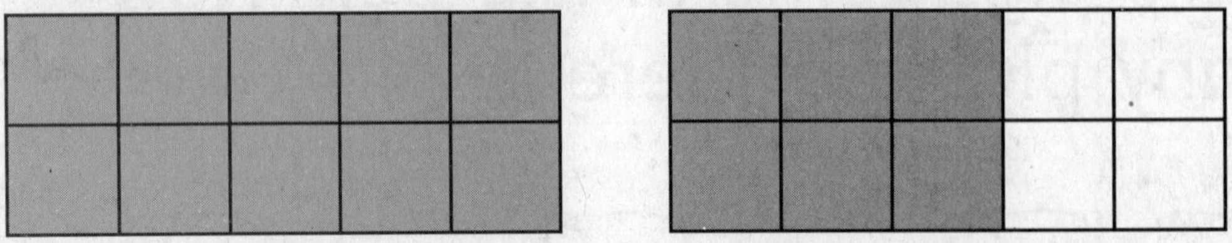

________ and ________ is ________

5. Count the number in each group. Write how many in all. K.OA.1

________ + ________ = ________ in all

6. Circle the pies to show K.OA.1

$2 + 6 = 8$

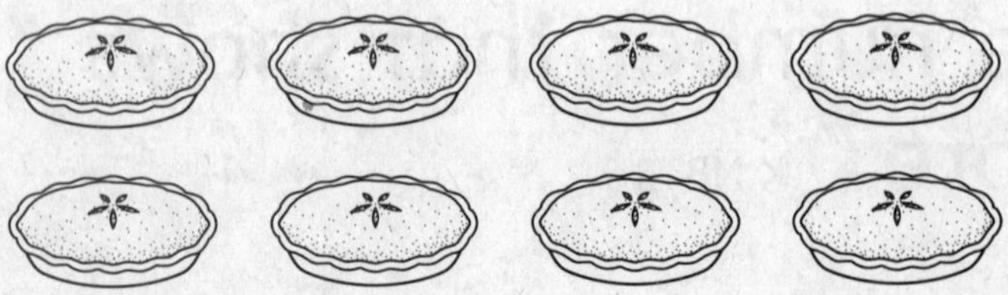

7. Count the number of each boat. Write each number. K.NBT.1

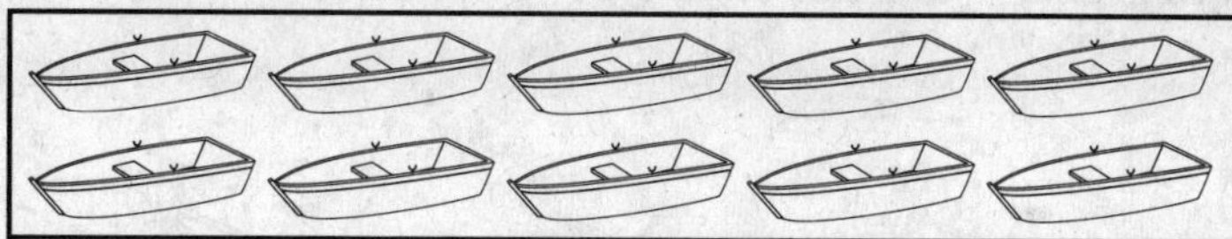	
______	______

8. Compare the objects. Circle the one that is shorter. Underline the one that is longer. K.MD.2

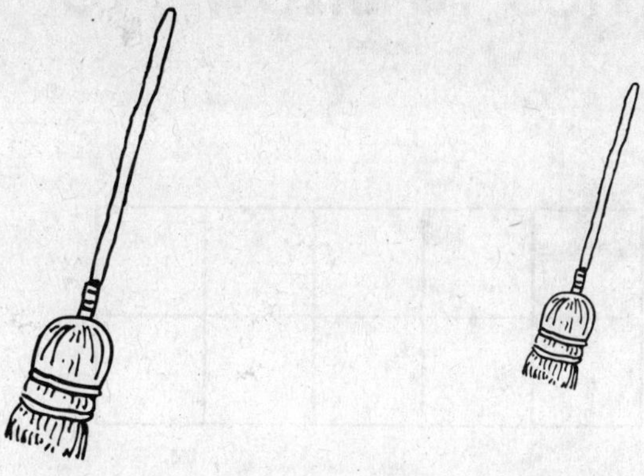

9. Underline the plus sign. Circle the subtraction sign. K.OA.1

× − + =

10. There are ten bikes. Six bikes ride away. Cross out six bikes. Write how many bikes are left. K.OA.3

11. Shade the ten frames to show "10 and 8 is 18." K.NBT.1

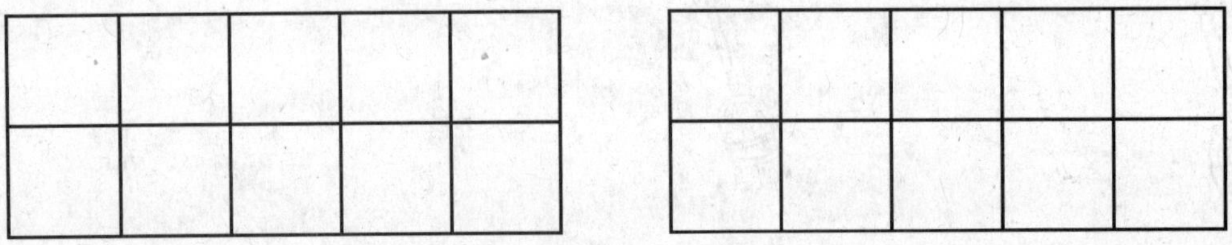

12. Circle the things that are too heavy to lift. K.MD.2

13. Count the number of bees. Count how many are away from the hive. Count how many are left in the hive. Write the numbers. K.OA.1

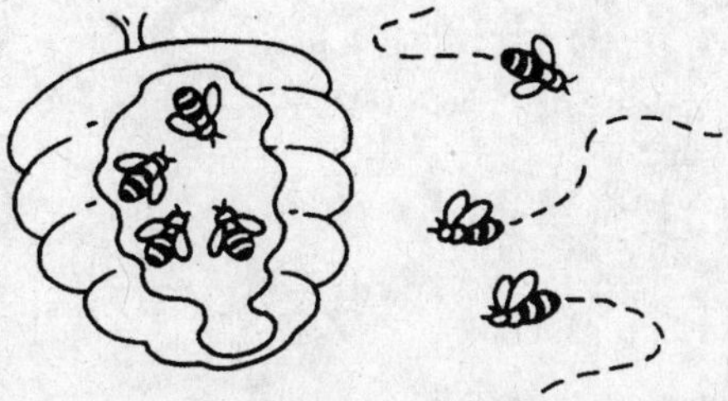

________ − ________ = ________

14. Circle the container that holds the most. K.MD.2

15. Henry has 4 balls. He loses one ball. How many balls does Henry have left? Circle the number. K.OA.1

1 2 3

16. Circle the heavier object. K.MD.2

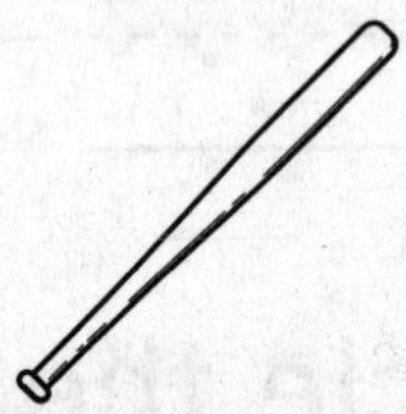

17. There are 3 dry dogs and 3 wet dogs. How many dogs are there in all? Circle the correct number. K.OA.1

5 6 7

18. How many pennies long is the dart? K.MD.2

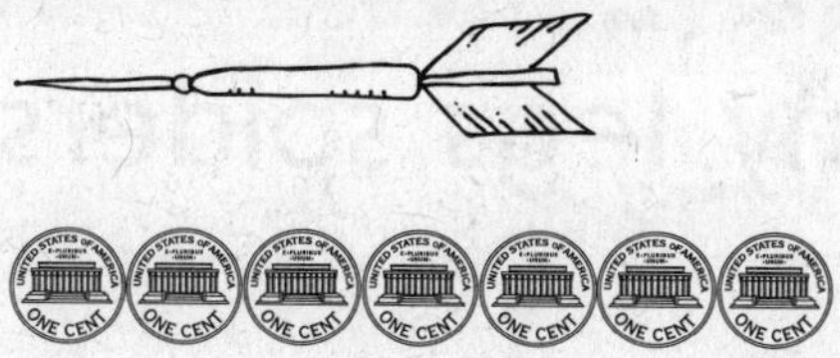

19. Cross out blocks to show the number sentence. K.OA.1

$$10 - 2 = 8$$

20. There are thirteen books on a shelf. Three books fell off the shelf. Cross out three books. How many books are left on the shelf? K.NBT.1

Performance Task

Studying Bugs

Research how many legs spiders and insects have.

Write your answers on another piece of paper. Show all your work to receive full credit.

K.CC.1, K.CC.3, K.OA.1

Part A

Draw a spider and an insect with the right number of legs. How many legs in all?

Spider Insect ______

K.CC.1, K.OA.1

Part B

There are three insects. Two more insects come. How many insects in all? Write a number sentence.

______ + ______ = ______

in all

NAME .... DATE ....

Benchmark Test 3

SCORE ....

1. Write the number that is one more than 4. K.CC.2, K.CC.4c

2. Count the apples. Write the number. Count the bananas. Write the number. Circle the number that is greater than the other. K.CC.5, K.CC.7

3. Circle the shapes you can roll. K.G.B.4

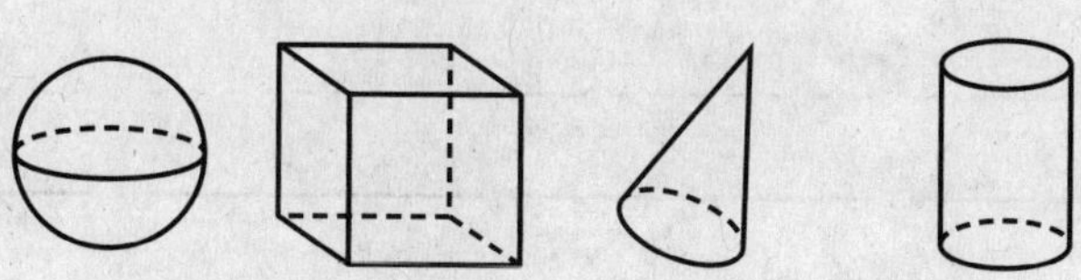

4. Color the boxes in the ten-frames to show 14. K.CC.4, K.CC.5

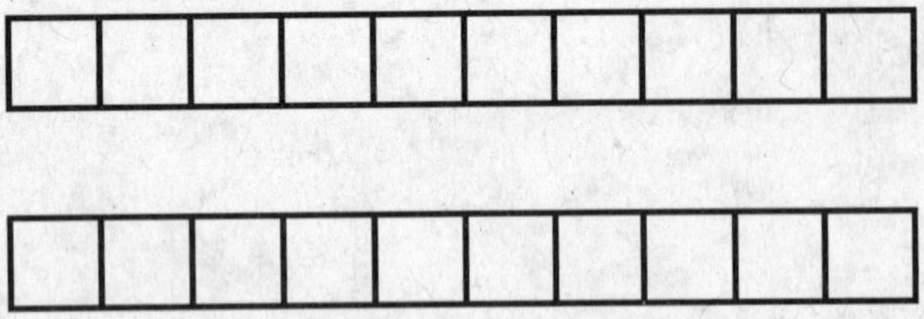

5. Count the tennis balls. How many are there? Write the number. K.CC.3, K.CC.5

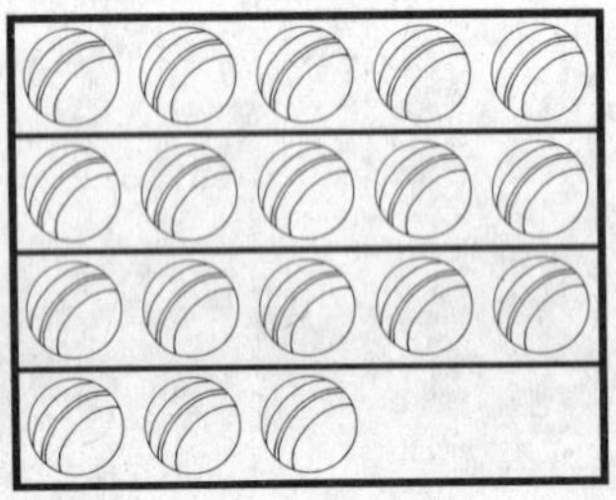

6. Count by 10s. Write the numbers. K.CC.1

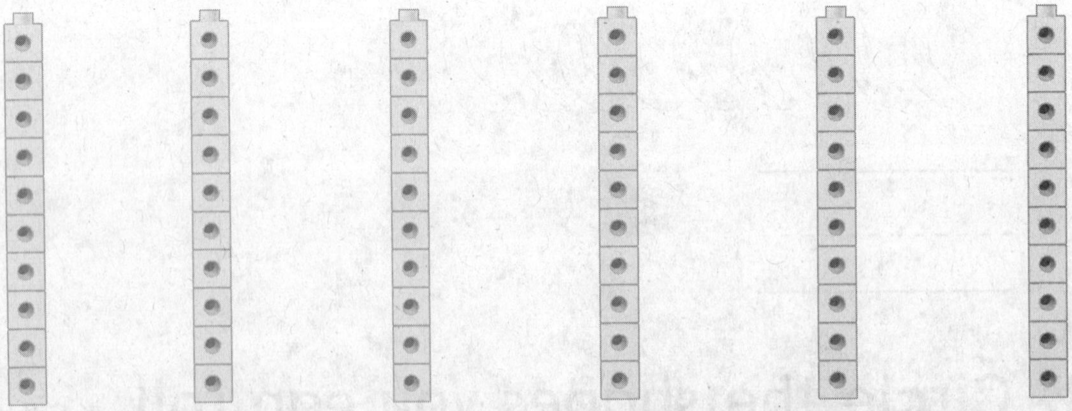

7. Circle the numbers that make five. K.OA.1, K.OA.3

2 and 3   2 and 4

1 and 4   1 and 3

8. What number does 3 and 1 make? Draw objects to show the number. Write the number. K.OA.2, K.OA.5

9. Look at the first shape in the row. Say the name of the shape to yourself. Compare this shape to the other shapes. Circle the matching shape. K.G.2, K.G.4

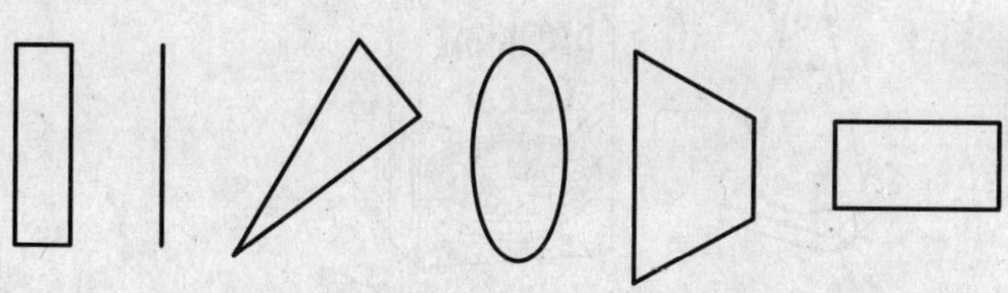

10. Write one way to make seven. Write another way to make seven. K.OA.1, K.OA.3

________ and ________ make 7

________ and ________ make 7

11. Count each group of crayons. Write the numbers. Write how many crayons there are in all. K.OA.1, K.OA.2

________ and ________ make ________

12. Circle the cylinder. K.G.4

13. Circle eight cars. Write the number of cars not circled to make 10. K.OA.4

8 and ________ make 10

14. Draw two groups of fish to make four fish in all. Write the number in each group. K.OA.1, K.OA.3

________	________

15. Draw an X through the number of hats that shows how many you would take away. K.OA.2

$6 - 2 = 4$

16. Match each word to the correct symbol. K.OA.1

equal sign	+
plus sign	−
minus sign	=

17. Eight apples are in the tree. You pick three. Draw an X on each apple you picked. How many are left? Write the number. K.OA.1, K.OA.2

______

18. Circle the group of objects that shows ten. Circle the group that shows some more. Write the number in each group. Write the number made. K.CC.4, K.NBT.1

______ and ______ make ______

19. Draw objects in the ten-frames to make sixteen. K.OA.1, K.NBT.1

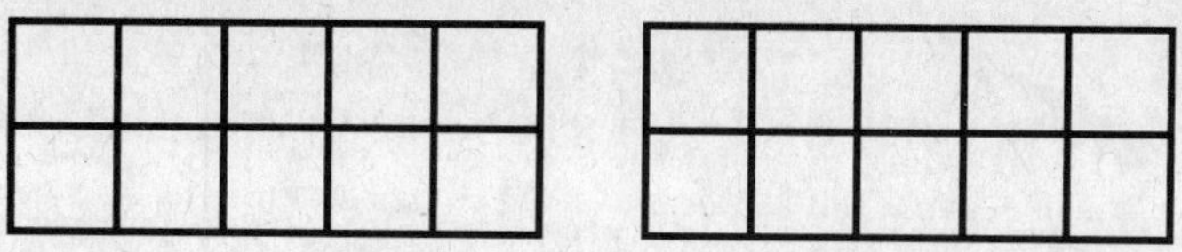

20. Count the number of objects in each ten-frame. Write each number. Trace the number made. K.CC.4, K.NBT.1

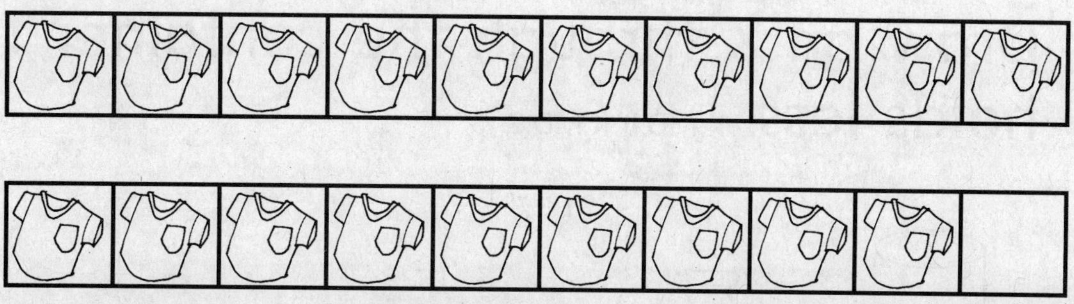

______ and ______ more make 19

21. Circle the train that is shorter. Draw an X through the train that is longer. K.MD.I, K.MD.2

22. Circle the animal that is lighter. Draw an X through the animal that is heavier. K.MD.I, K.MD.2

23. Circle the container that holds more. Draw an X through the container that holds less. K.MD.I, K.MD.2

24. Circle the dog that is different. Draw an X through the dogs that are the same. K.MD.3

25. Circle the shape that is different. K.MD.3

26. Circle each apple. Draw an X through each banana. Draw a box around each pear. K.MD.3

27. Circle the two-dimensional shapes. Draw an X through the three-dimensional shapes. K.G.B.3

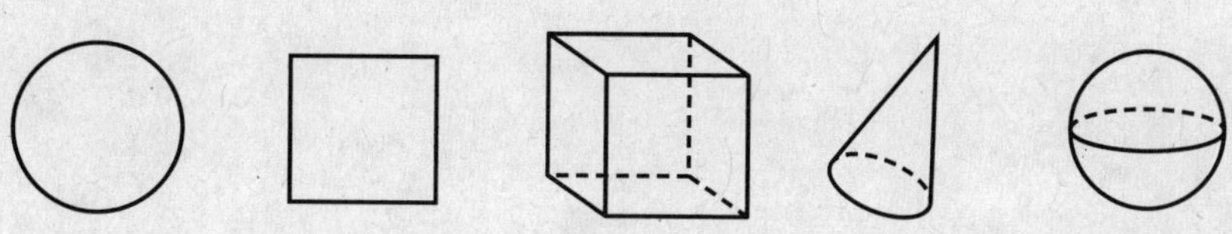

28. Draw a balloon above the table. Draw a box under the table. K.G.1

29. Circle the object beside the flower. K.G.1

30. Draw a triangle. K.G.5

31. Circle the shape that has four sides. K.G.4

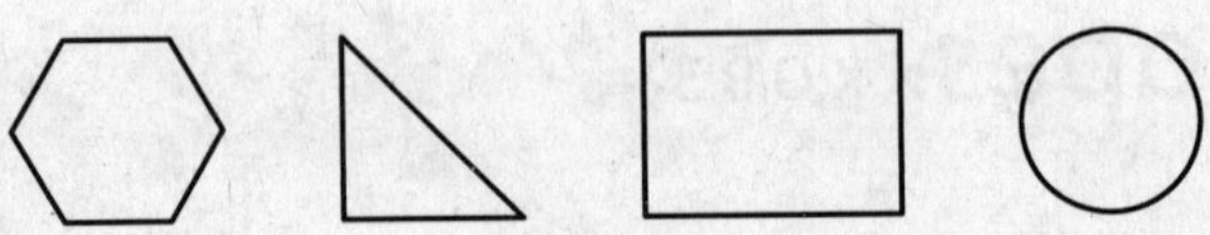

Performance Task

Packing Tomatoes

Farmer Laura is packing tomatoes in boxes before selling them.

Part A

How many tomatoes are in each of these two boxes? Write the number.

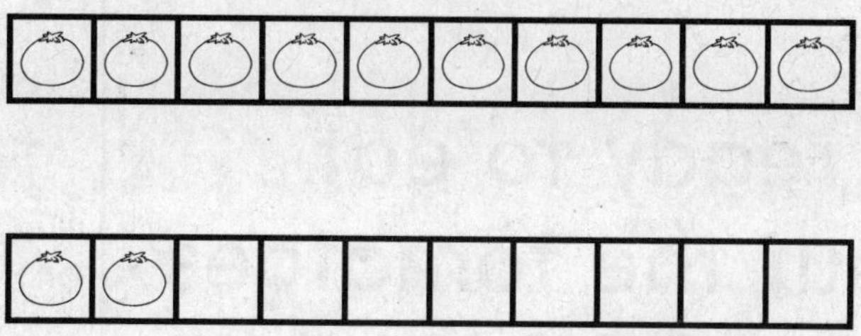

Part B

How many tomatoes are there in all? Count the tomatoes and write the number.

Part C

Look at the box of tomatoes. How many tomatoes do you need to fill the box and make 10? Write the number.

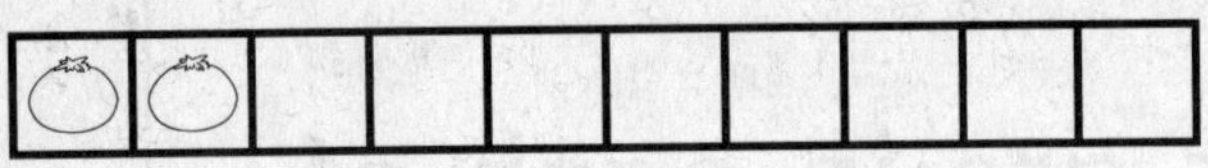

Part D

These tomatoes are ready to eat. You eat two. Cross out the tomatoes you eat. How many are left? Write the number.

NAME ______________________ DATE ______________________

Benchmark Test 4

SCORE ______________________

1. Look at the first shape in the row. Say the name of the shape to yourself. Compare this shape to the other shapes. Circle the matching shape. K.G.2, K.G.4

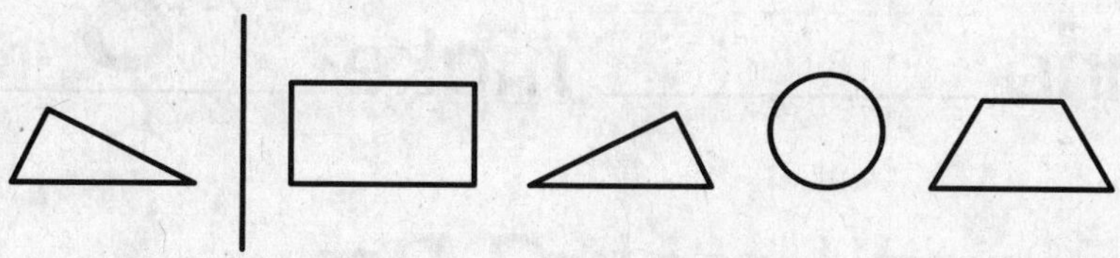

2. Circle the shovel that is shorter. Draw and X through the shovel that is longer. K.MD.I, K.MD.2

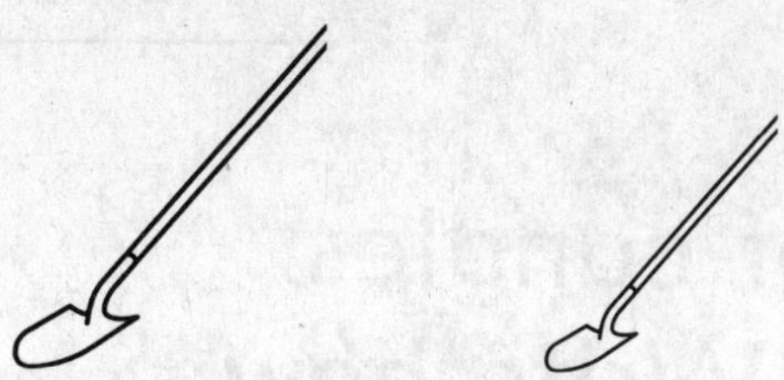

3. Circle the cupcake that is different. Draw an X through the cupcakes that are the same. K.MD.3

4. Write one way to make eight. Write another way to make eight. K.OA.1, K.OA.3

____ and ____ make 8

____ and ____ make 8

5. What do 5 and 1 make? Draw circles to show the number. Write the number. K.OA.2, K.OA.5

____

6. Count each group of candles. Write the numbers. Write how many candles there are in all. K.OA.1, K.OA.2

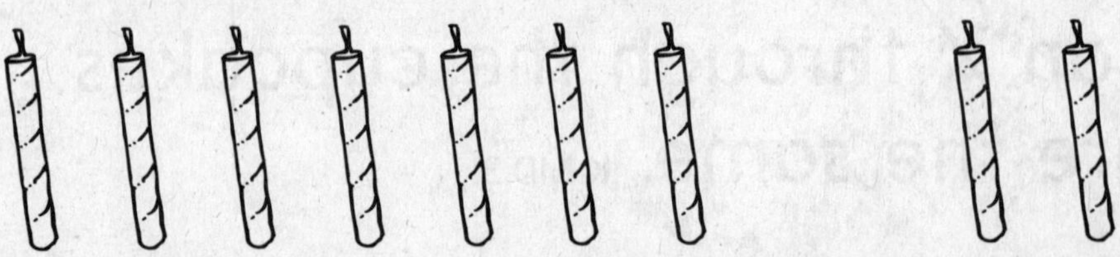

____ and ____ make ____

7. Draw objects in the ten-frame to show 14. K.OA.1, K.NBT.1

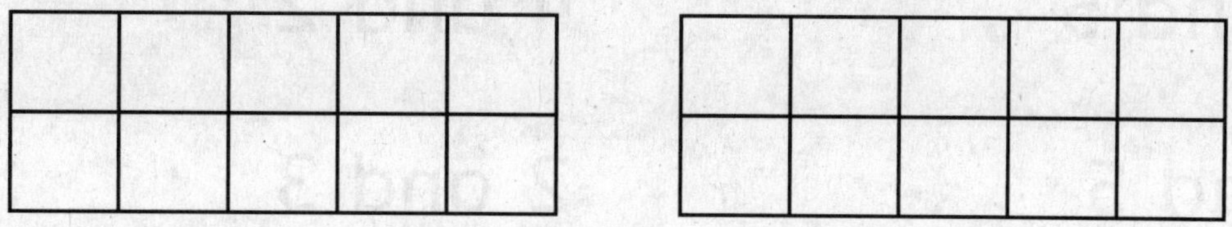

8. Circle the group of objects that shows ten. Circle the group that shows some more. Write the number in each group. Write the number made. K.CC.4, K.NBT.1

_______ and _______ make _______

9. Count the cars. Write the number. Count the airplanes. Write the number. Circle the number that is greater than the other. K.CC.5, K.CC.7

10. Circle the numbers that make 5. K.OA.1, K.OA.3

0 and 5 | 4 and 2
1 and 5 | 2 and 3

11. Nine desks are in a classroom. The teacher removes three. Draw an X on each desk that is removed. How many are left? Write the number. K.OA.1, K.OA.2

______

12. Circle the animal that is lighter. Draw an X through the animal that is heavier. K.MD.1, K.MD.2

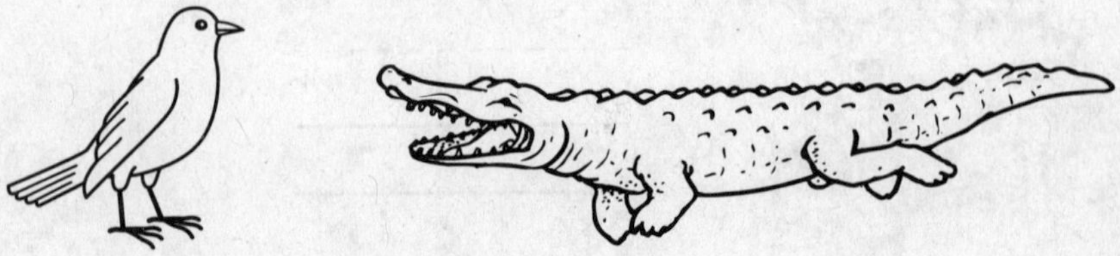

13. Circle the container that holds more. Draw an X through the container that holds less. K.MD.1, K.MD.2

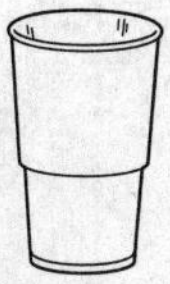

14. Write the number that is one more than 3. K.CC.2, K.CC.4c

____

15. Circle each dog. Draw a box around each cat. Draw an X through each bird. K.MD.3

16. Draw an X through the number of sleds that shows how many you would take away. K.OA.2

8 − 3 = 5

17. Count the number of objects in each ten-frame. Write each number. Write the number made. K.CC.4, K.NBT.1

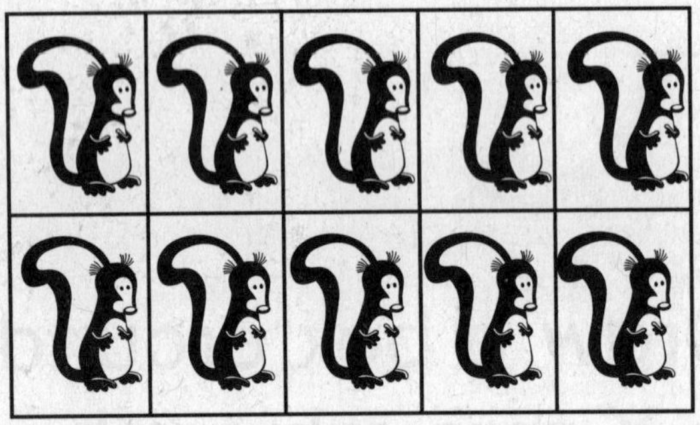

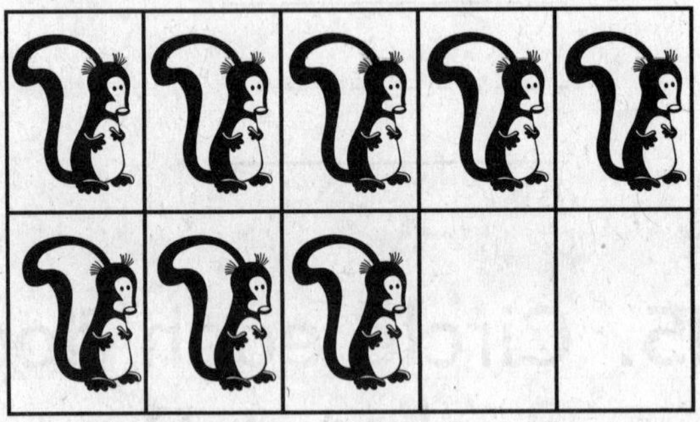

__________ and __________ make __________

18. Circle the shape that has three sides. K.G.4

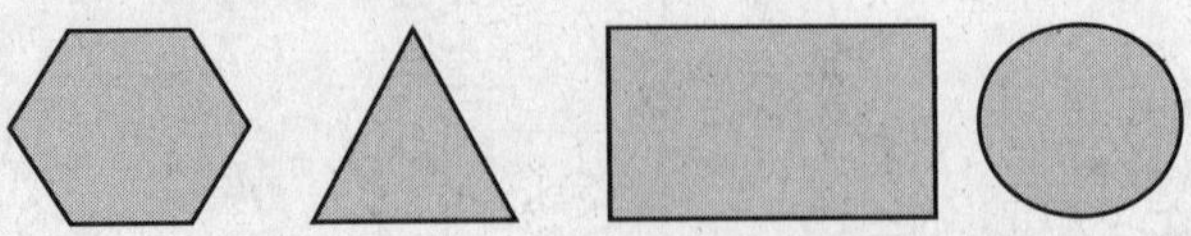

19. Count the baseball bats. How many are there? Write the number. K.CC.4, K.CC.5

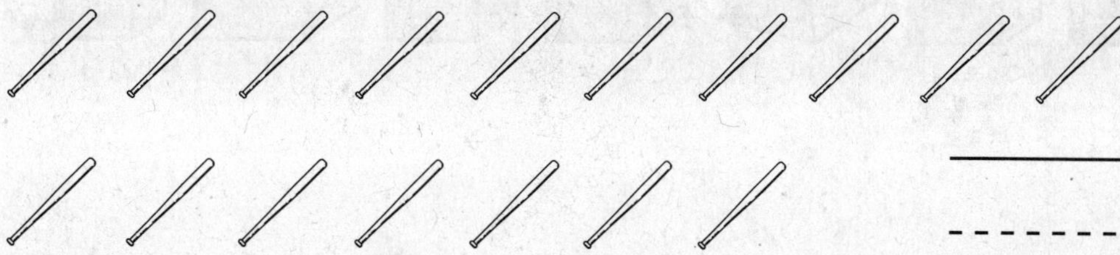

______

20. Count by 10s. Write the numbers. K.CC.1

10

______

______

______

21. Match each word to the correct symbol. K.OA.1

equal sign	–
plus sign	+
minus sign	=

22. Circle the shape that is different. K.MD.3

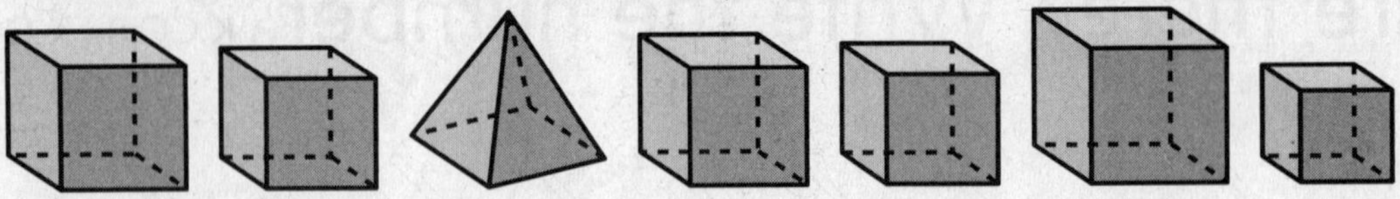

23. Draw two groups of smiley faces to make five smiley faces in all. Write the number in each group. K.OA.1, K.OA.3

__________ and __________

24. Circle the three-dimensional shapes. Draw an X through the two-dimensional shapes. K.G.3

25. Circle the shapes you can stack. K.G.4

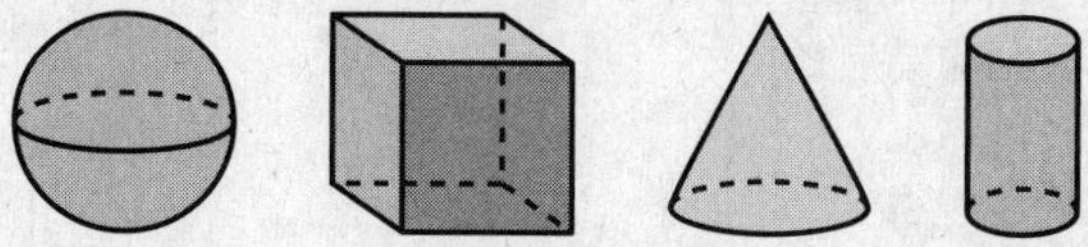

26. Circle seven beds. Write the number of beds not circled to make 10. K.OA.4

7 and ________ make 10

27. Draw a bird above the house. Draw a worm below the house. K.G.1

28. Circle the cone. K.G.4

29. Color the boxes in the ten-frames to show 13. K.CC.3, K.CC.5

30. Circle the object beside the hammer. K.G.1

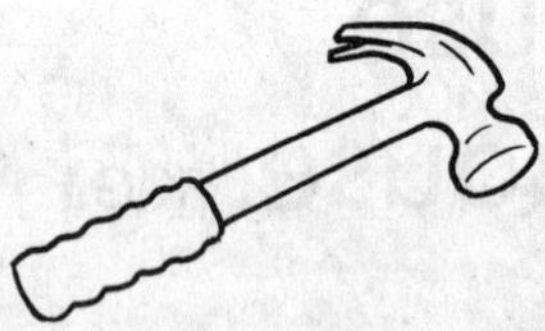
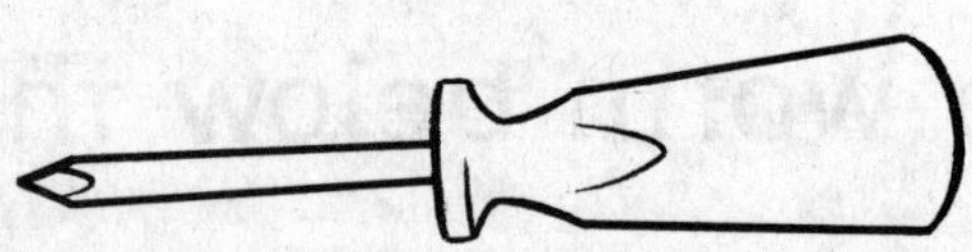
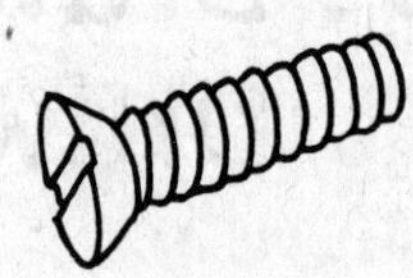

31. Draw a square. K.G.5

Performance Task

Stacking Books

A librarian is stacking books on shelves.

Write your answers on another piece of paper. Show all your work to receive full credit.

Part A

K.CC.3, K.CC.4b, K.CC.5

How many books are on each of these two shelves? Write the number.

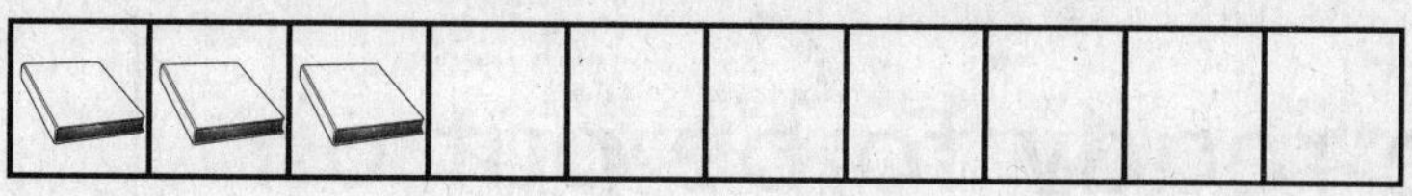

Part B

K.CC.3, K.CC.5, K.OA.1, K.OA.2, K.NBT.1

How many books are there in all? Count the books and write the number.

Performance Task *(continued)*

K.CC.3, K.OA.1, K.OA.2, K.OA.4

Part C

Look at the shelf of books. How many books do you need to fill the shelf and make 10? Write the number.

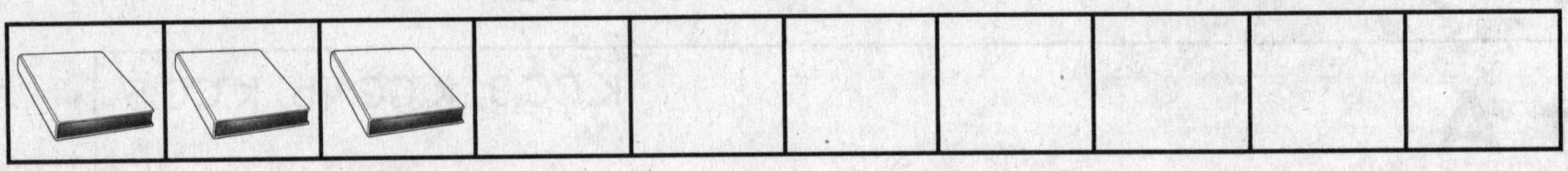

K.CC.3, K.OA.1, K.OA.2

Part D

These books are ready to be put on a shelf. The librarian puts two on a shelf. Cross out the books she puts on the shelf. How many are left? Write the number.

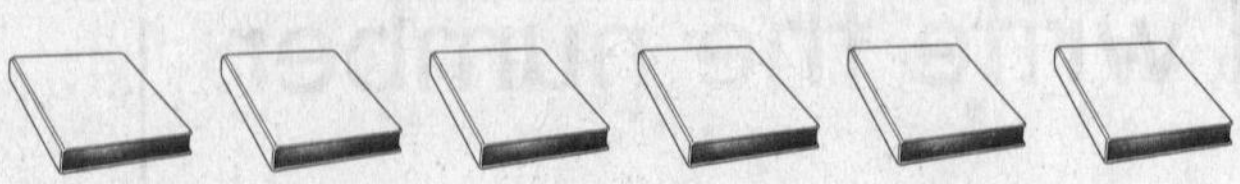

Smarter Balanced Assessment Item Types

In Grade 3, students will begin to take the Smarter Balanced Test for Math. It is a test that is taken on a computer. The problems on the next few pages show the kinds of questions students might have to answer and what to do to show the answer on the computer. However, the student materials in this book center mainly on ***Multiple Choice*** and ***Click to Select*** formats.

Selected Response means that you are given answers from which you can choose.

Selected Response Items

Regular multiple choice questions are like tests you may have taken before. Read the question and then choose the one best answer.

Multiple Choice

Sasha has a collection of coins. There are 3 in one pile and 2 in another. How many coins does she have in all?

- ☐ 2
- ☐ 3
- ☐ 5

ONLINE EXPERIENCE Click on the box to select the one correct answer.

HELPFUL HINT Only one answer is correct. You may be able to rule out some of the answer choices because they are unreasonable.

Sometimes a multiple choice question may have more than one answer that is correct. The question may or may not tell you how many to choose.

Multiple Correct Answers

Choose **all** the statements that tell the story.

- ☐ $2 + 3 = 5$
- ☐ $5 = 2 + 3$
- ☐ $10 - 2 = 8$

ONLINE EXPERIENCE Click on the box to select it.

HELPFUL HINT Read each answer choice carefully. There may be more than one right answer.

Another type of question asks you to tell whether the sentence given is true or false. It may also ask you whether you agree with the statement, or if it is true. Then you select yes or no to tell whether you agree.

Multiple True/False or Multiple Yes/No

There were no birds on the pond. Two brown ducks swim into the pond. Five geese fly in and land on the pond. Tell whether you agree with each sentence.

Yes	No	
☐	☐	There are more geese than ducks on the pond.
☐	☐	There are 2 birds on the pond.
☐	☐	5 geese flew away.
☐	☐	There are 7 birds in all on the pond.

ONLINE EXPERIENCE Click on the box to select it.

HELPFUL HINT There is more than one statement. Any or all of them may be correct.

You may have to choose your answer from a group of objects.

Click to Select

Which shape is like a car tire?

ONLINE EXPERIENCE Click on the figure to select it.

HELPFUL HINT On this page you can draw a circle or a box around the figure you want to choose.

Sometimes you must use your mouse to click on an object and drag it to the correct place to create your answer.

Drag and Drop

Move these objects so that they match the story.

ONLINE EXPERIENCE You will click on an object and drag it to the spot it belongs.

HELPFUL HINT Either draw a line to show where the object goes or sketch the object in the right place.

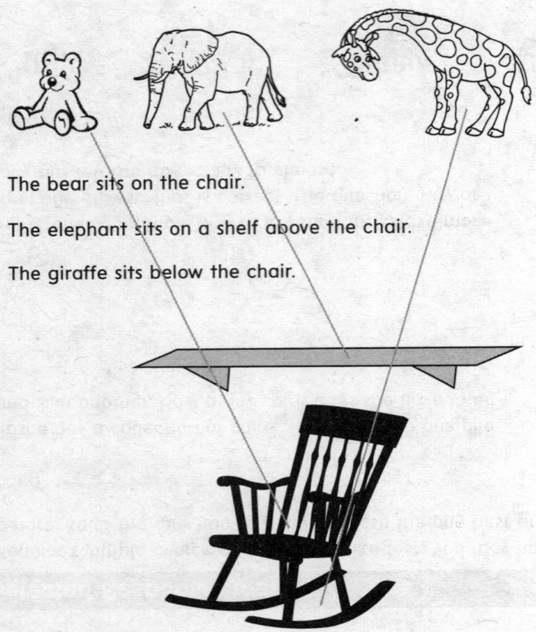

The bear sits on the chair.

The elephant sits on a shelf above the chair.

The giraffe sits below the chair.

When no choices are given from which you can choose, you must create the correct answer. One way is to type in the correct answer. Another may be to make the correct answer from parts that are given to you.

Constructed-Response Items

Fill in the Blank

Loren has 3 fish in her tank. For her birthday she will receive 2 more fish. Write a number sentence to show how many fish she will have after her birthday.

ONLINE EXPERIENCE You will click on the space and a keyboard will appear for you to use to write the numbers and symbols you need.

HELPFUL HINT There is more than one way to write the correct sentence.

3 (+) 2 = 5

NAME ______________ DATE ______________

SCORE ______________

Countdown: 20 Weeks

1. Identify the number of circles. K.CC.3

Draw four circles.

Students should draw four circles.

Write the numeral that shows the number of circles you drew.

4

2. Count the soccer balls. Circle the number that tells how many. K.CC.4a

1 (2) 3

3. Circle the group with more dogs. K.CC.4a

4. Circle the group with three flowers. K.CC.4a

5. Circle the bowl with zero oranges. K.CC.4a

NAME ____________________ DATE ____________

SCORE ____________

Countdown: 19 Weeks

1. Write the number of umbrellas shown. K.CC.3

5

2. Count the spoons. Find the group of forks with the same number. Draw a line from the group of spoons to the group of forks with the same number. K.CC.4a

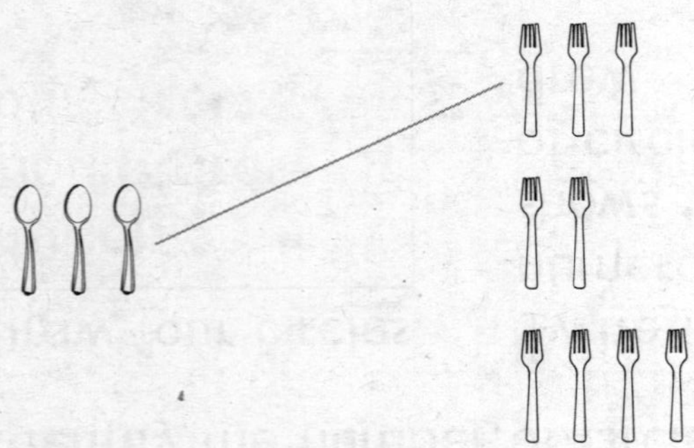

3. Draw an X on the groups that show less than 4. K.CC.4a

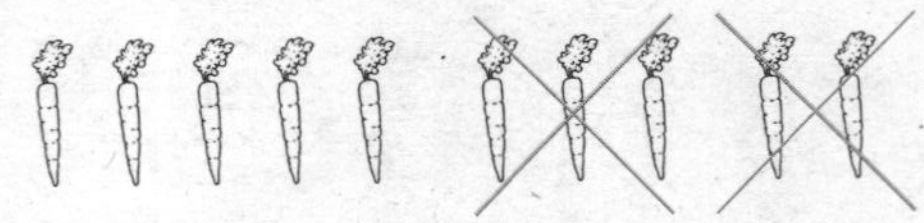

4. Circle the number that is one more than two. K.CC.6

2 (3) 1

5. Draw a group of bugs that is less than the group shown. K.CC.4a

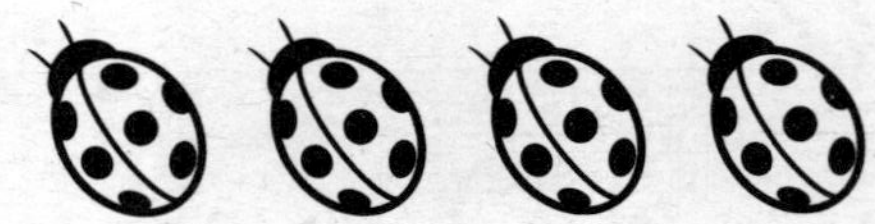

Students should draw a group of 1, 2, or 3 bugs.

NAME ________________ DATE ________________

SCORE ________________

Countdown: 18 Weeks

1. Count and draw lines to match objects in one group with objects in the other. Write the numbers. Circle the number and group that is greater than the other. K.CC.3

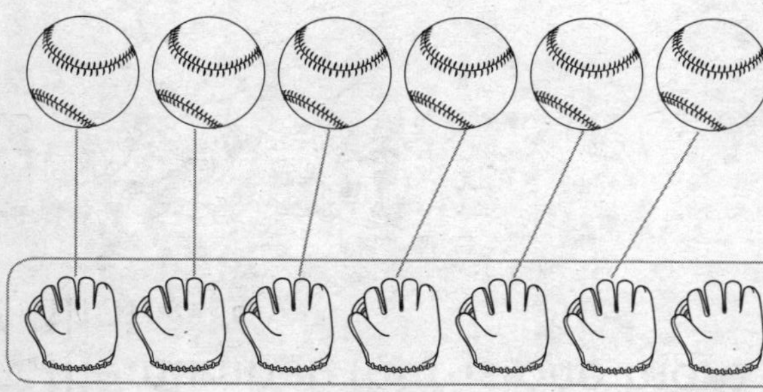

6

7

2. Draw a box around the second rock from the frog. Draw an X through the third rock. Draw a line under the first rock. K.CC.4a

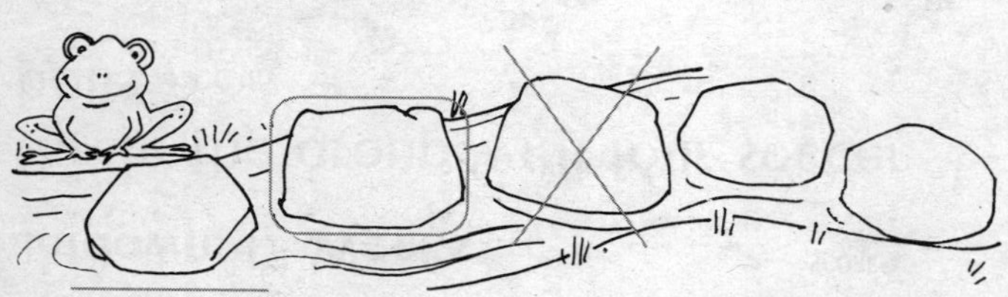

3. Count the houses. Draw one more house. Count how many. Write the number. K.CC.3

9

4. Circle the number that is one more than one. K.CC.4c

0 2 1 3

5. Count the violins. Circle the number that tells how many. K.CC.4b

7 8 6

NAME ________________ DATE ________________

Countdown: 17 Weeks

SCORE ________________

1. Circle the group with four soccer balls. K.CC.4a

2. Count the number of animals. Circle the number that is one more than that. K.CC.4c

5 4 3

3. Color the boxes in the ten-frame to show eight. K.CC.1

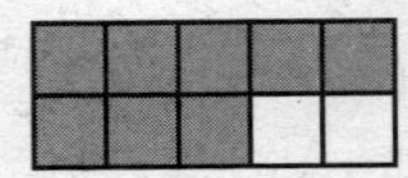

4. Circle the number that is one more than two. K.CC.4c

0 2 1 3

5. Count the pumpkins. Write the number that tells how many. K.CC.3

9

Copyright © McGraw-Hill Education. Permission is granted to reproduce for classroom use.

NAME ____________ DATE ____________

Countdown: 16 Weeks

SCORE ____________

1. Count the giraffes. Write how many. K.CC.3

12

2. Circle 15 geese. K.CC.3

3. Draw more circles to make 19. K.CC.3

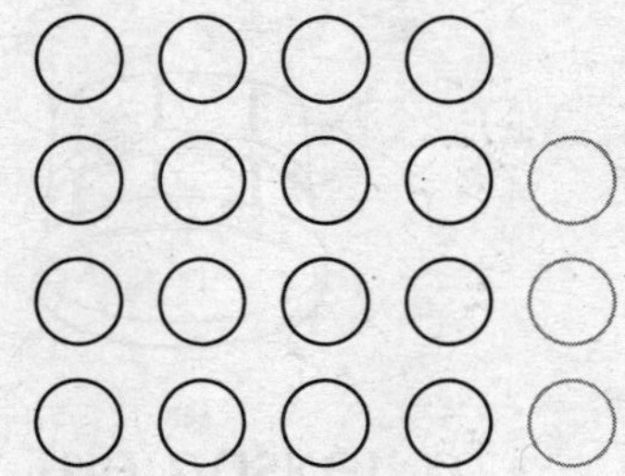

4. Connect the numbers in order. K.CC.3

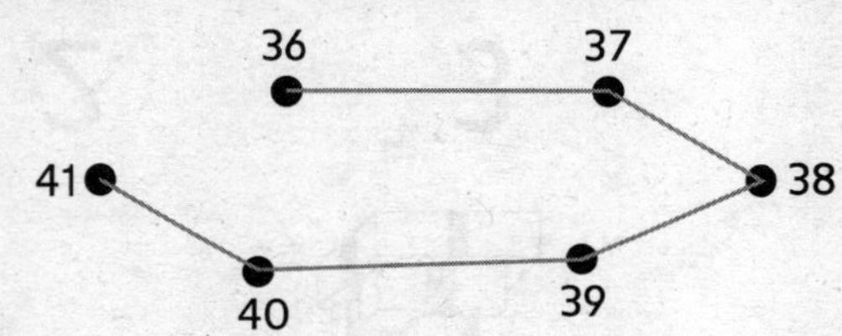

5. Circle the number that is one more than 1. K.CC.4c

1 0 2

Countdown

NAME ____________ DATE ____________

SCORE ____________

Countdown: 15 Weeks

1. Count the pieces of pasta. Write one more than the number. K.CC.3

11

2. Count and draw lines to match objects in one group with objects in the other. Write the numbers. Circle the group that has more than the other. K.CC.3

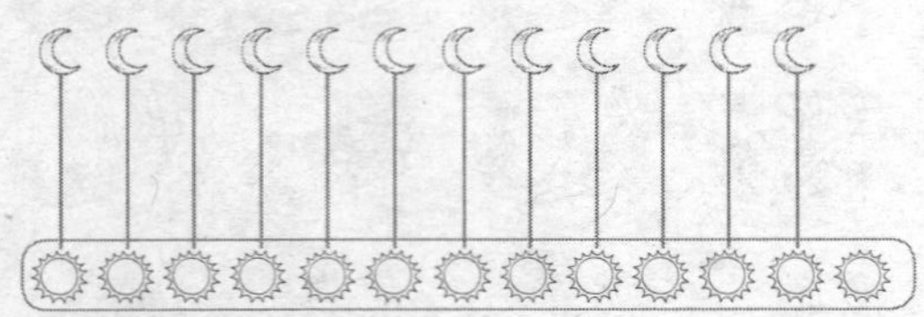

12

13

3. Color more boxes in the ten-frame to show 18. K.CC.1

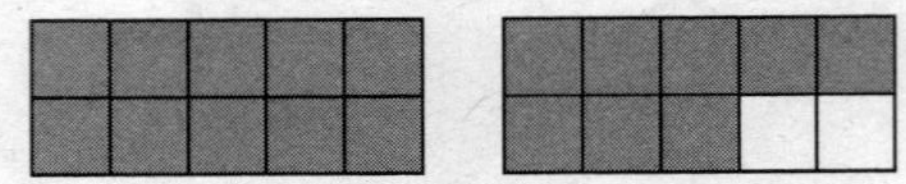

4. Count the presents. Circle the number that is one more than that. K.CC.4c

2 3 4

5. Circle the fourth strawberry from the basket. K.CC.4a

NAME ______________ DATE ______________

SCORE ______________

Countdown: 14 Weeks

1. Count by tens. Write the number. K.CC.3

10 20 30 40

2. There are 4 cars. Take apart the number. K.OA.1

2 and 2

3. Count how many. Write the number beside the carrots. Take apart the number. K.OA.3

5

1 and 4

4. There are 6 cats. Circle cats to make 4 cats and 2 cats. K.OA.7

5. There are 7 chipmunks. Take apart the number. K.OA.1

4 and 3

Sample answers given. Also correct: 0 and 7, 1 and 6, 2 and 5, 3 and 4, 5 and 2, 6 and 1, 7 and 0.

Countdown

NAME DATE

SCORE

Countdown: 13 Weeks

1. Count the balloons. Write the number beside the balloons. K.OA.1

8

Show a way to take apart the number.

Sample answer:

5 and 3

2. There are 9 squares. Break apart a group of 6. Write the number in the other group. K.OA.3

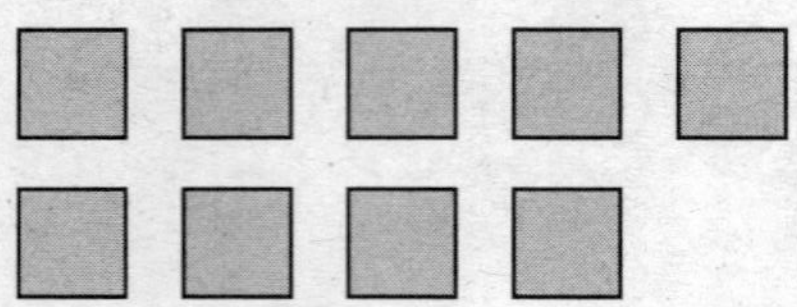

9

6 and 3

3. Take apart 10. Write the numbers that make 10. K.OA.4

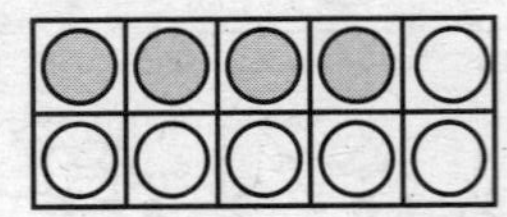

4 and 6

4. Continue counting by tens. K.CC.3

60 70 80 90

5. Circle the group with 6 pineapples. K.CC.3

NAME ________________ DATE ________________

SCORE ________________

Countdown: 12 Weeks

1. There are 10 crabs. Take apart the crabs into 2 groups. Write the numbers. K.OA.3 Sample answers given. Also correct: 0 and 10, 1 and 9, 2 and 8, 3 and 7, 4 and 6, 5 and 5, 6 and 4, 7 and 3, 9 and 1, 10 and 0.

8 and 2

2. There are two crayons on the table. Joey puts two more crayons on the table. How many crayons are there in all? K.OA.1

4

3. Write the number that tells how many there are in all. K.OA.1

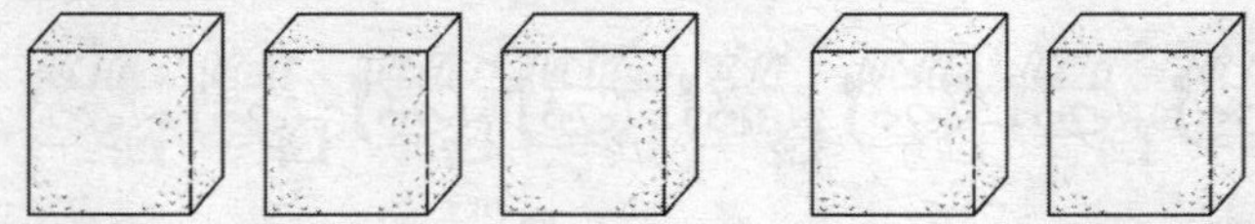

3 + 2 = 5 in all

4. Count the number in each group. Write the number. Write how many in all. K.OA.1

4 + 3 = 7 in all

5. Circle two groups to match the number sentence. K.OA.1

$2 + 5 = 7$

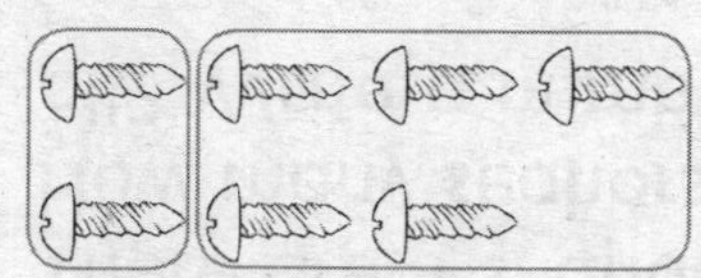

Countdown

NAME .... DATE ....

SCORE ....

Countdown: 11 Weeks

1. There are 10 hats. Take apart the hats into 2 groups. Write the numbers. K.OA.3

Sample answers are given. Also correct: 0 and 10, 1 and 9, 2 and 8, 3 and 7, 4 and 6, 6 and 4, 7 and 3, 8 and 2, 9 and 1, 10 and 0.

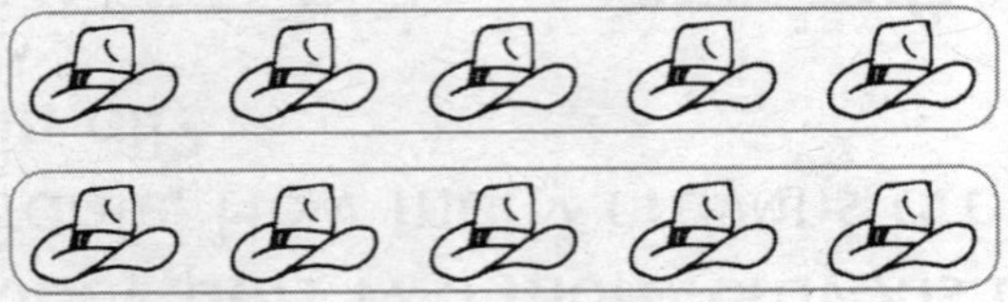

5 and 5

2. Count the number in each group. Write the numbers. Then write the number in all. K.OA.1

3 + 6 = 9

3. There are 4 seahorses. 1 more comes. How many seahorses are there in all? Write a number sentence. K.OA.1

4 + 1 = 5

4. Count the objects and write the numbers. K.OA.4

3 + 7 = 10

5. Count the cows. Write one more than that many. K.CC.3

9

Copyright © McGraw-Hill Education. Permission is granted to reproduce for classroom use.

Copyright © McGraw-Hill Education. Permission is granted to reproduce for classroom use.

NAME ____________ DATE ____________

SCORE ____________

Countdown: 10 Weeks

1. Count the number in each group. Write the numbers. Then write the number in all. K.OA.1

6 + 1 = 7

2. Kumar had 3 shovels. He took one away. Cross out one shovel. Write how many are left. K.OA.1

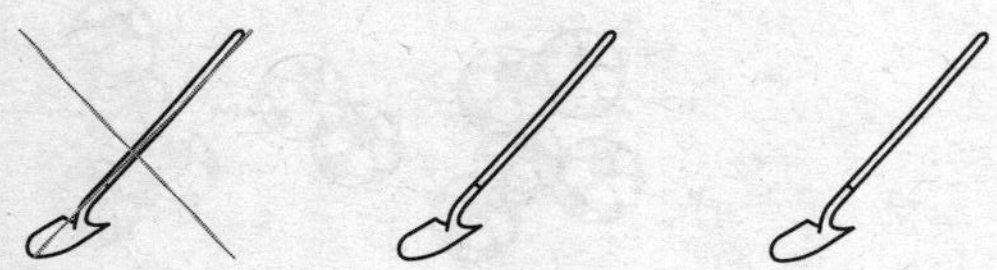

2

3. There are 7 pieces of pizza. Enrique eats 2 pieces. Cross out two pieces. Write how many pieces are left. K.OA.1

5

4. Count the number of snakes. Count how many are going away. Count how many are left. Write the numbers. K.OA.1

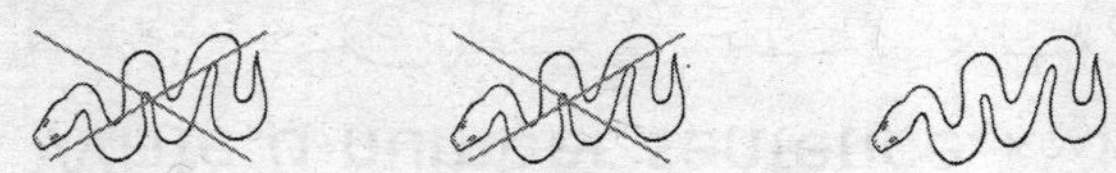

3 – 2 = 1

5. Write a number sentence. K.OA.1

7 – 7 = 0

Countdown

NAME ______________ DATE ______________

Countdown: 9 Weeks

SCORE ______________

1. Count the soccer balls. Three are kicked away. Cross out three soccer balls. Write how many are left. K.OA.1

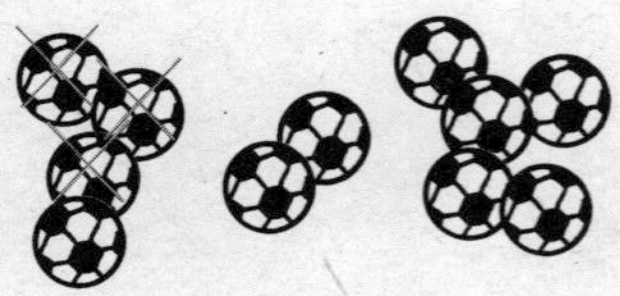

8

2. There are 8 spiders. Four spiders crawl away. How many are left? Write a number sentence. K.OA.1

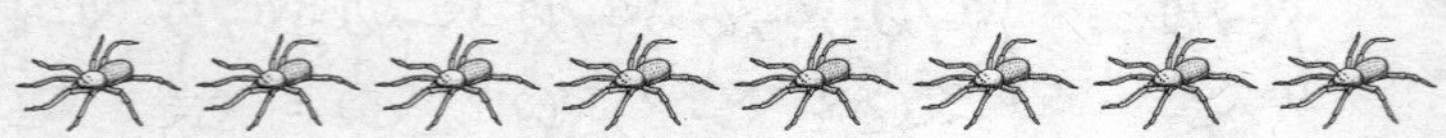

8 – 4 = 4

3. There are ten starfish. Eight starfish go back in the ocean. Cross out eight starfish. Write how many starfish are left. K.OA.3

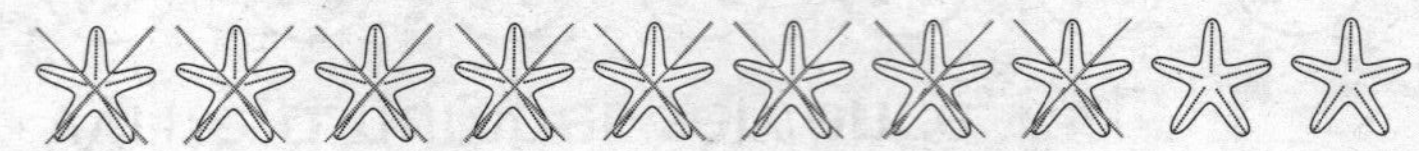

2

4. Write a number sentence. K.OA.1

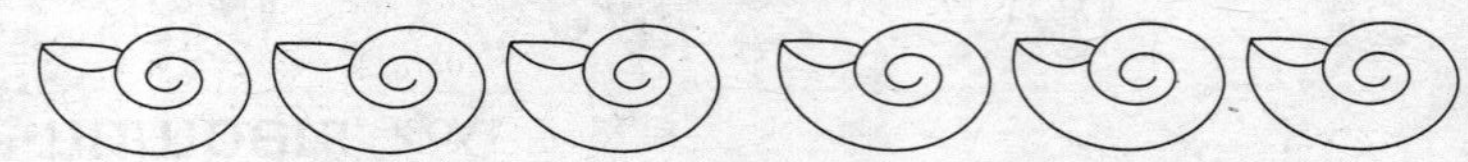

3 + 3 = 6

5. Count the objects and write the numbers. K.OA.4

1 + 9 = 10

Copyright © McGraw-Hill Education. Permission is granted to reproduce for classroom use.

NAME ______________________ DATE ______________

SCORE ______________

Countdown: 8 Weeks

1. Draw an X through the strawberries to show how many you would take away. K.OA.7

$6 - 4 = 2$

2. Count the numbers in each ten-frame. Write the numbers. Write how many in all. K.NBT.1

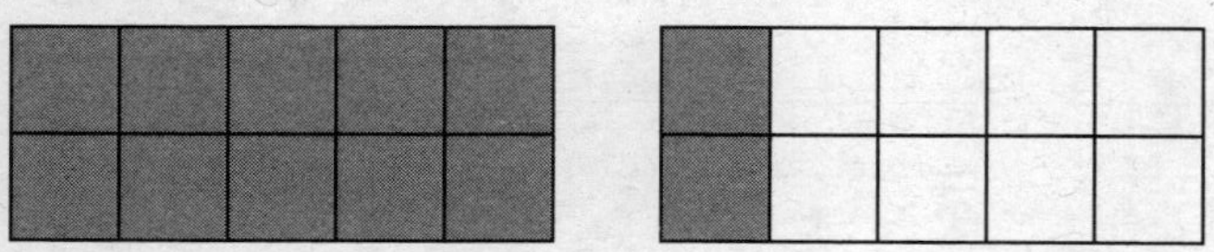

10 and 2 is 12

3. Shade the ten-frames to show "10 and 3 is 13." K.NBT.1

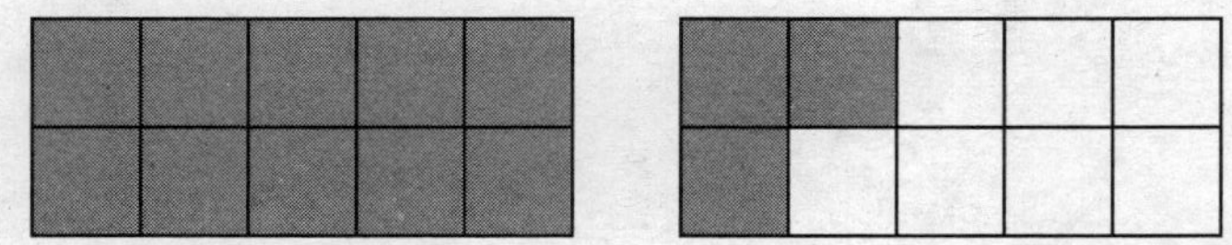

4. Count the number of each present. Write each number. K.NBT.1

10	4

5. Circle the answer to "ten and one more." K.NBT.1

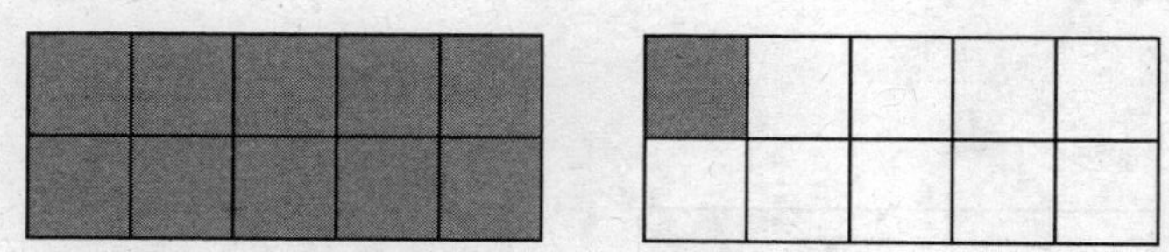

10 (11) 12

NAME ______ DATE ______

SCORE ______

Countdown: 7 Weeks

1. Match each word to the correct symbol. K.OA.1

minus sign — +

plus sign — −

equal sign — =

2. Circle the picture that shows 10 − 3. K.OA.3

3. Shade the ten-frames to show "10 and 5 more." Write the total number. K.NBT.1

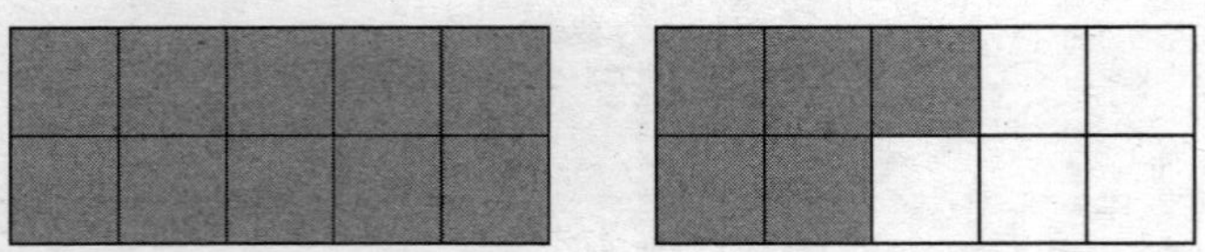

15

4. Which is correct? K.NBT.1

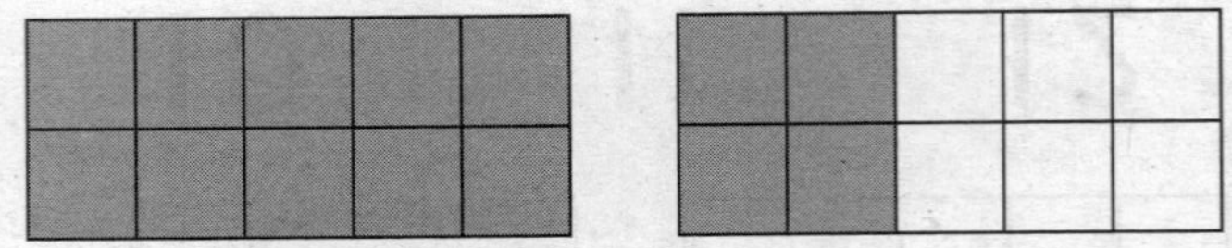

10 and 2 more

10 and 3 more

10 and 4 more

5. Count the number of each animal. Write each number. K.NBT.1

NAME ______________________ DATE ______________

SCORE ______________

Countdown: 6 Weeks

1. Count the numbers in each ten-frame. Write the numbers. Write how many in all. K.NBT.1

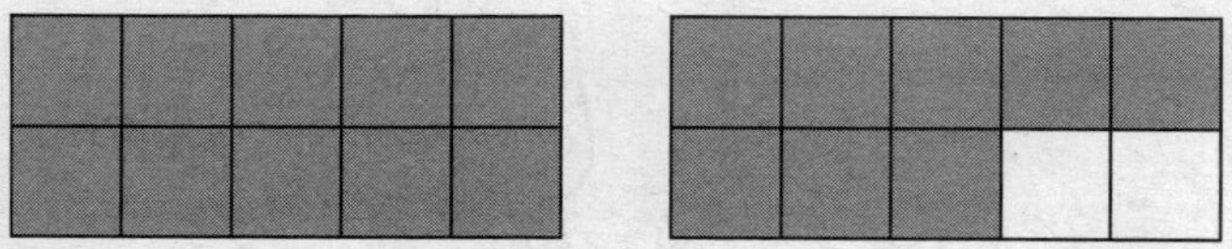

10 and 8 is 18

2. Compare the objects. Circle the one that is longer. Put an X through the one that is shorter. K.MD.2

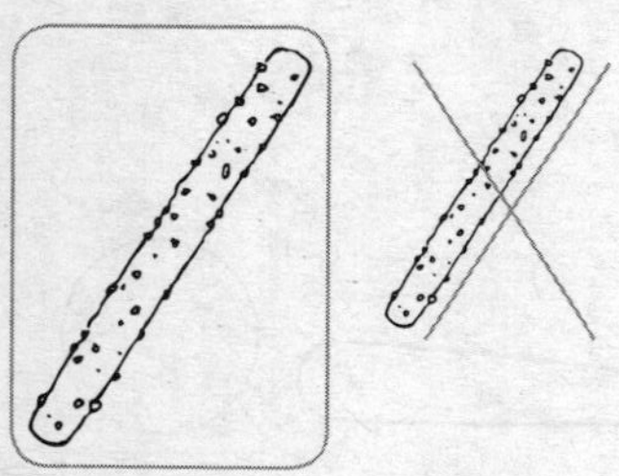

3. Compare the giraffes. Circle the taller one. Underline the shorter one. K.MD.2

4. Circle the things that are too heavy to lift. K.MD.2

5. Look at the two objects. Circle the object that is taller and weighs more than the barrel. K.MD.1

NAME ____________ DATE ____________

SCORE ____________

Countdown: 5 Weeks

1. Circle the ones that are alike. K.MD.3

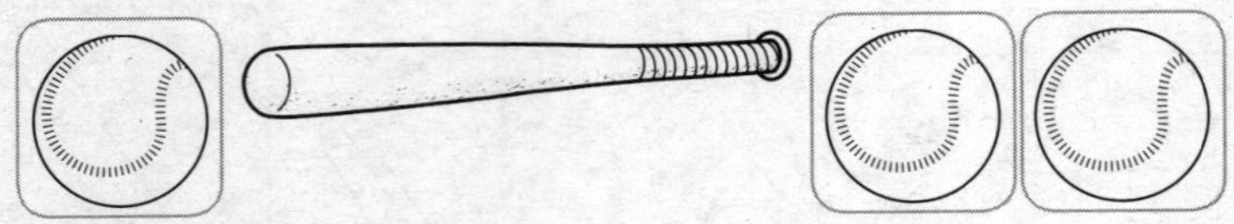

2. Look at the pictures. Circle the pictures that do not belong in a house. K.MD.3

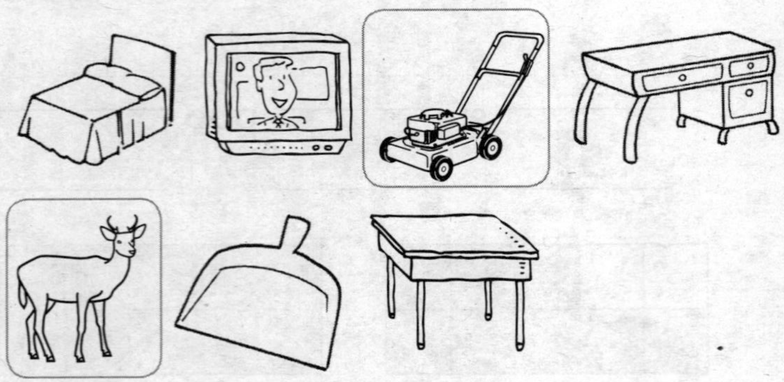

3. Sort by size. Draw X's through the bigger ones. K.MD.3

4. Sort by shape. Circle all the ones that are alike. K.MD.3

5. Match the object to the word. K.MD.2

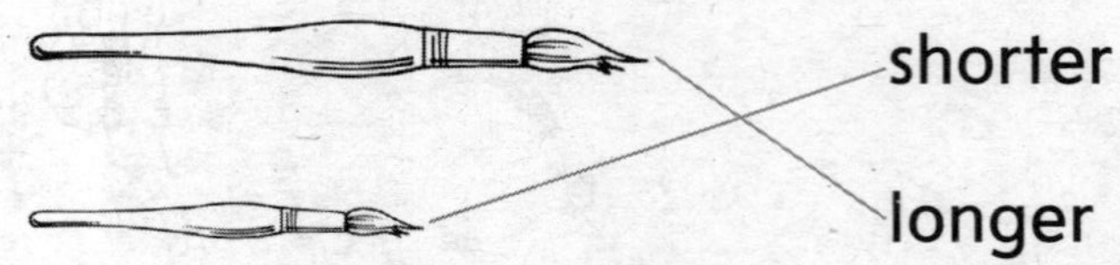

NAME ______________________ DATE ____________

SCORE ____________

Countdown: 4 Weeks

1. Circle the heavier object. Cross out the lighter object. K.MD.2

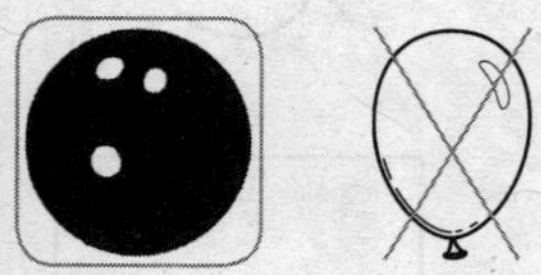

2. Circle the object that is above the log. K.G.1

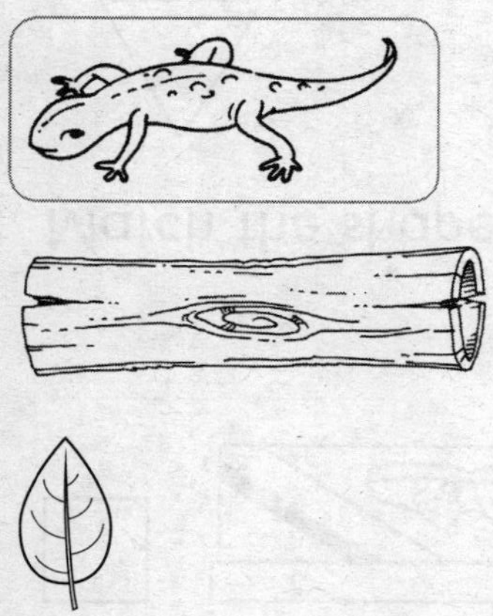

3. Circle the dog that is in front of the doghouse. K.G.1

4. Circle the animals that are next to the duck. K.G.1

5. Draw a circle beside the key. K.OA.7

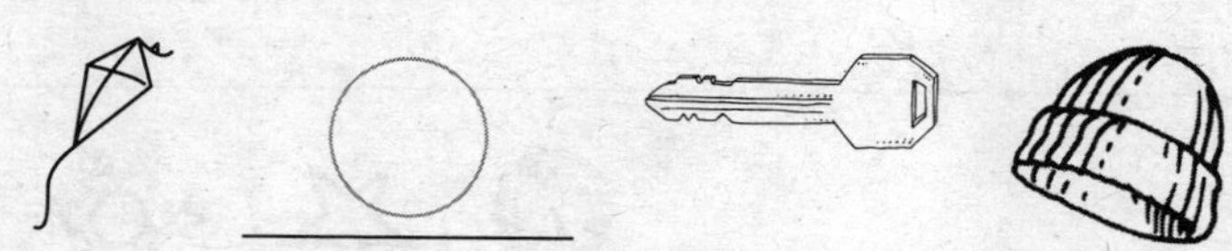

NAME ______________________ DATE ______________

SCORE ______________

Countdown: 3 Weeks

1. Put an X on the squares. K.G.2

2. Match the shape to the name. K.G.4

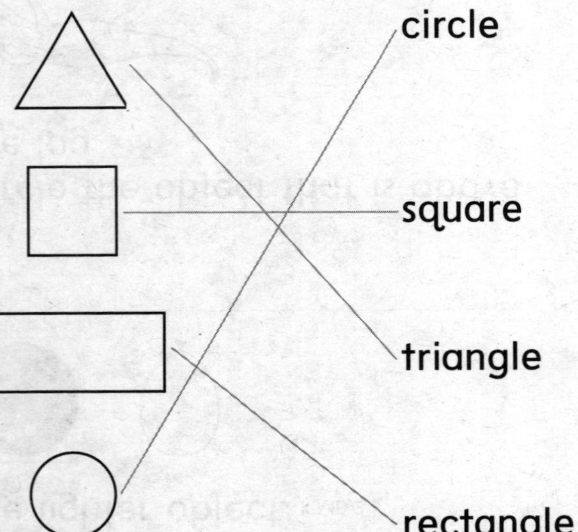

3. Underline the object that has a hexagon. K.G.2

4. Draw the shape that comes next. K.G.4

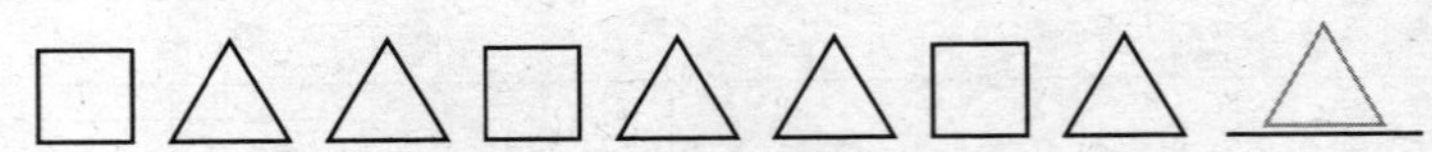

5. Circle the fifth ball from the bat. Underline the first ball. Cross out the second ball. K.CC.4a

NAME ______________________ DATE ______________

SCORE ______________

Countdown: 2 Weeks

1. Circle the one that is different. K.MD.3

2. Circle the shapes that are used to make the design. K.G.5

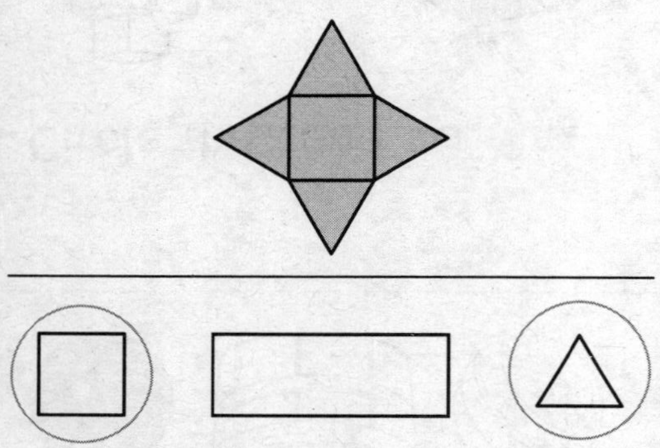

3. Draw the shape that comes next. K.G.4

4. Cross out the pictures that do not have circles. K.G.2

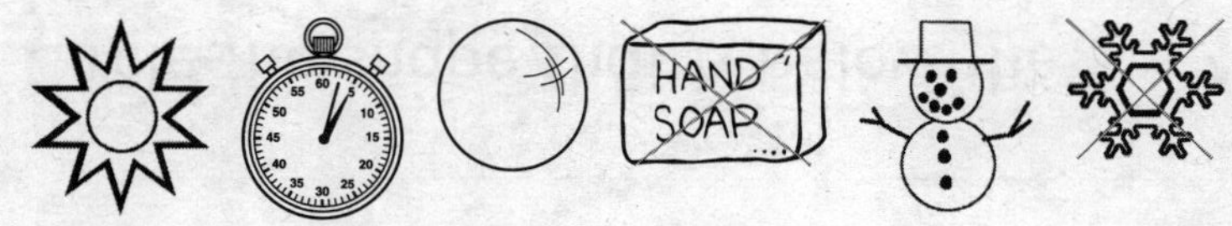

5. Underline the two objects that hold the same amount. K.MD.2

NAME ____________________ DATE ____________

SCORE ____________

Countdown: 1 Week

1. Put an X on the shapes that are not spheres. K.G.2

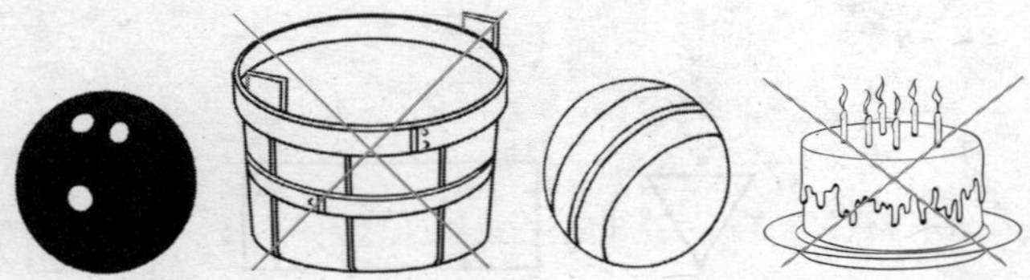

2. Circle the cone. K.G.1

3. Underline the shapes that can be stacked. K.G.4

4. Circle the shape that best matches the drum. K.OA.7

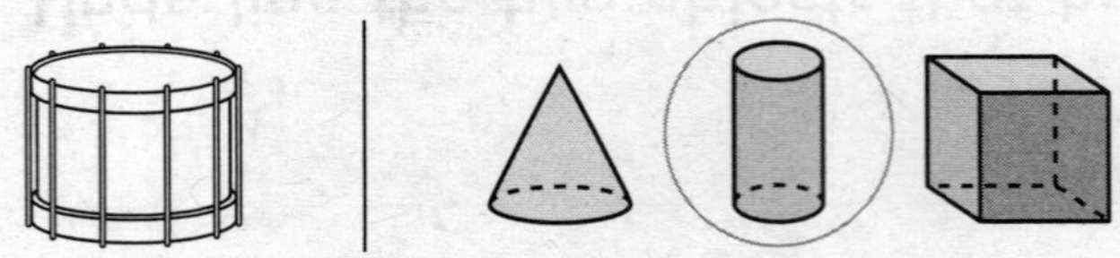

5. Circle the shape that is below the line. K.G.1

Copyright © McGraw-Hill Education. Permission is granted to reproduce for classroom use.

NAME ________ DATE ________

Chapter 1 Test

SCORE ________

1. Count the number of balloons. Circle the number that tells how many. K.CC.3, K.CC.5

1 (2) 3

2. Circle the bowl with zero peas. K.CC.6

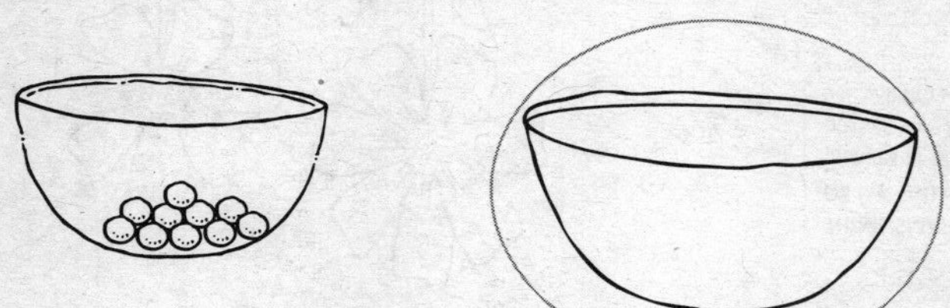

3. Draw three triangles. Write the numeral that tells the number of triangles. K.CC.4a, K.CC.3

Students should draw three triangles.

3

4. Circle the group with more peppers. K.CC.6

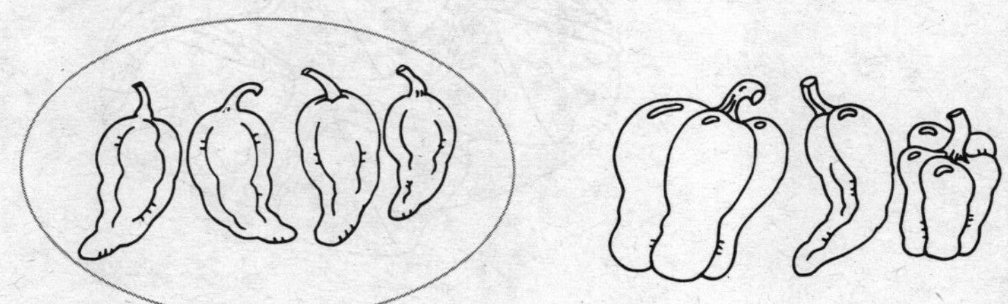

5. Draw a group of flowers that has more than the group shown. K.CC.4a, K.CC.6

THINK SMART FOR SBAC
On the test, you might drag pictures of flowers to an answer box. In this book, you will draw the flowers.

Students should draw more than 2 flowers.

6. Circle the group with five pencils. K.CC.4a, K.CC.1

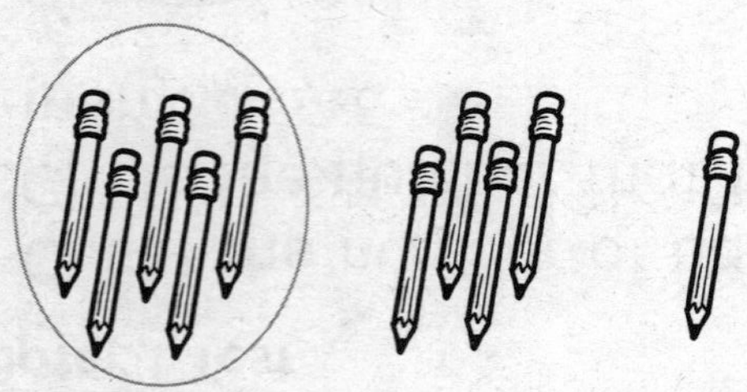

7. Write the number of sandwiches shown. K.CC.3, K.CC.5

4

8. Write the number of balloons shown. K.CC.3, K.CC.5

5

9. Draw an X on the groups that show less than 3. K.CC.5, K.CC.6

10. Circle the number that is one more than one. K.CC.4c

0 1 2 3

11. Count the dogs. Find the group of bones with the same number. Draw a line from the group of dogs to the group of bones with the same number. K.CC.4a, K.CC.6

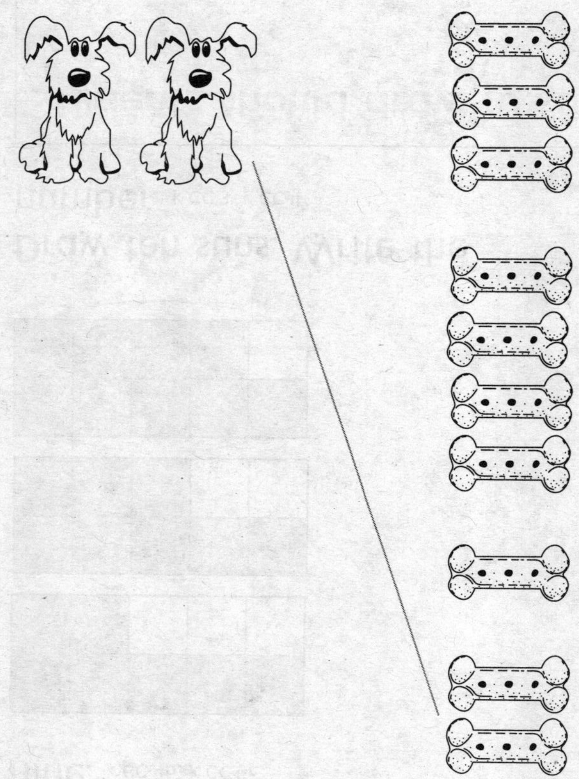

NAME ________________ DATE ________________

SCORE ________________

Chapter 2 Test

1. Color the boxes in the ten-frame to show seven. K.CC.4

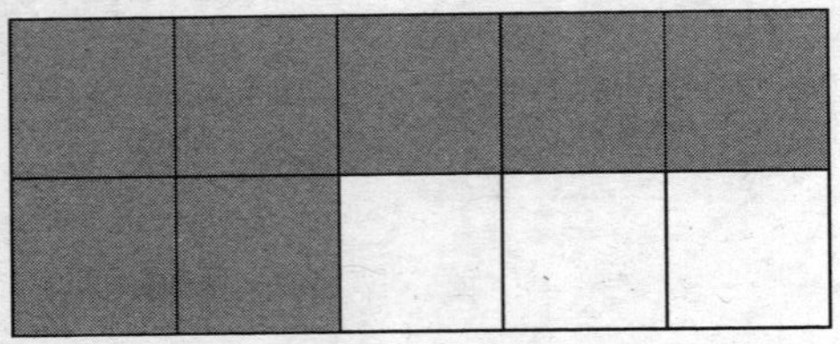

Students should fill in the top five boxes and the first two boxes on the left in the second row.

2. Count the circles. Draw more circles to show eight. Write the number. K.CC.3, K.CC.4

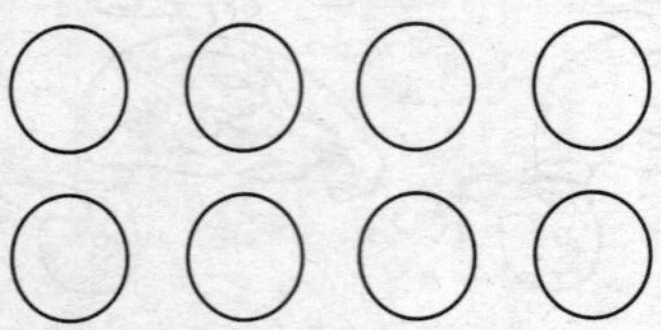

8

Students should draw 2 circles.

3. Circle the ten-frame that shows nine. K.CC.4b, K.CC.4c

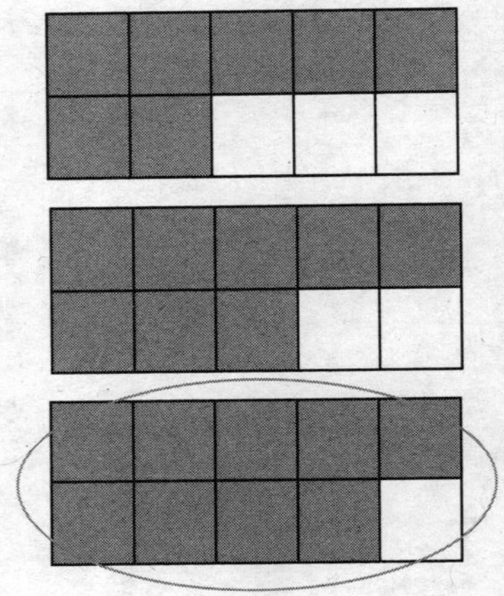

4. Draw ten suns. Write the number. K.CC.3, K.CC.4

Students should draw 10 suns.

10

Copyright © McGraw-Hill Education. Permission is granted to reproduce for classroom use.

Copyright © McGraw-Hill Education. Permission is granted to reproduce for classroom use.

5. Count the bunnies. Write the number. Count the carrots. Write the number. Circle the group and number that is greater than the other. K.CC.3, K.CC.6

7

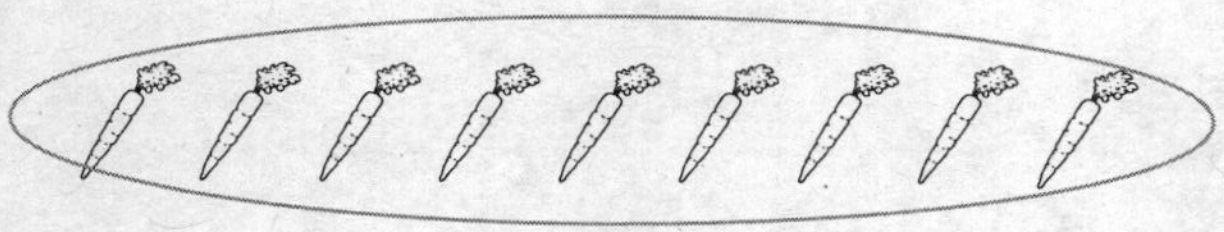

9

6. Show five and one more pencils on the ten-frame. Write the number of pencils. K.CC.3, K.CC.4

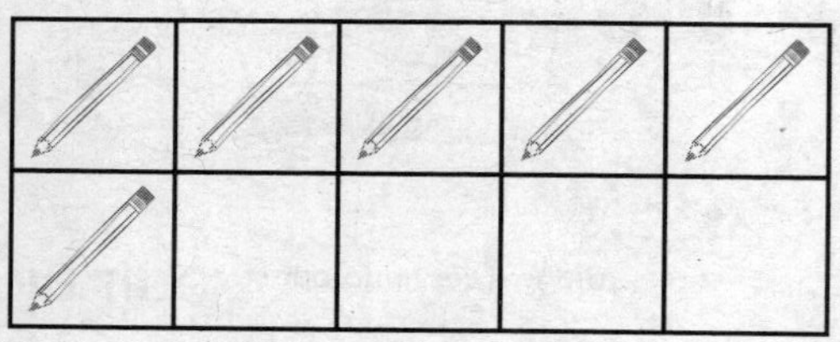

6

7. Write the number that matches the number of bears shown. K.CC.4a, K.CC.3

8

8. Circle the number that is one more than six. K.CC.4c

5 6 7 8

9. Circle the ant that is first in the line of ants leaving the picnic. Draw an X on the ant that is third in line. Draw a line under the ant that is fourth in line. Preparation for K.CC.4a

10. Circle the boat that is seventh in line from the lighthouse. Draw an X on the boat that is fifth in line. Draw a line under the boat that is tenth in line. Preparation for K.CC.4a

NAME ________________ DATE ________________

Chapter 3 Test

SCORE ________________

1. Count the ducks. Write how many. K.CC.3

11

2. Circle 16 dolls. K.CC.3

3. Draw more circles to make 12. K.CC.3

4. Connect the numbers in order. Circle the smallest number. K.CC.1

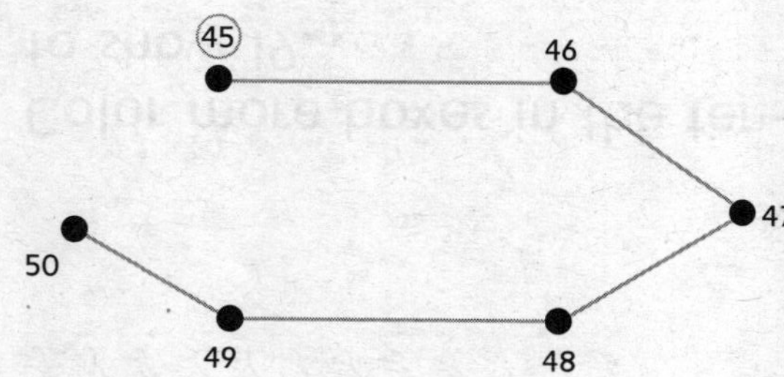

5. How many dinosaurs are there? Circle the correct number. K.CC.3

16 (17) 18

6. Continue counting by ones. K.CC.1

87 88 89 90

7. Count and draw lines to match objects in one group with objects in the other. Write the numbers. Circle the group that has more than the other. K.CC.3

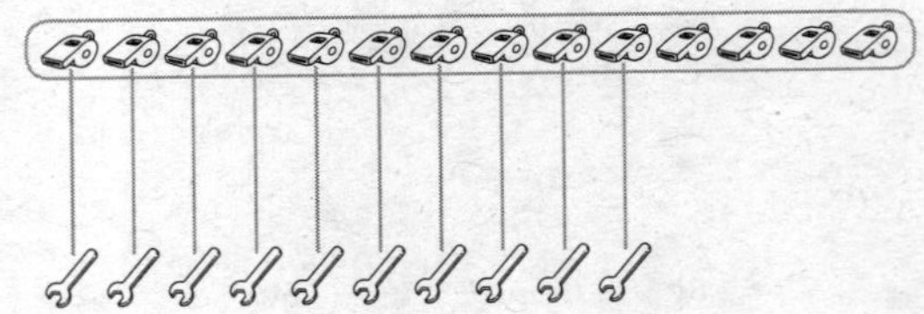

14

10

8. Color more boxes in the ten-frame to show 19. K.CC.3

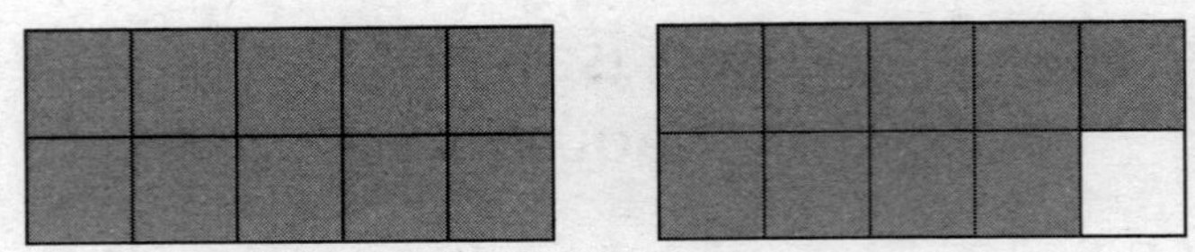

9. Count by tens. What number comes next? Write the number. K.NBT.1

30 40 50 60

10. Write the number shown by the ten-frames. K.CC.3

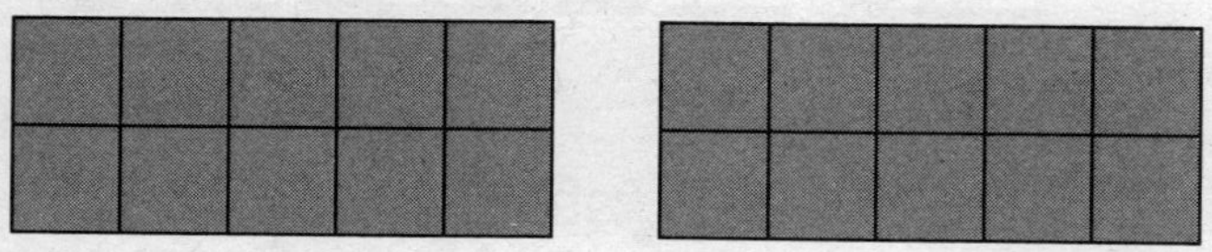

20

11. Circle the number that is one less than 16. K.CC.1

13 14 15

12. Count the watches. Circle the same number of boats. K.CC.1

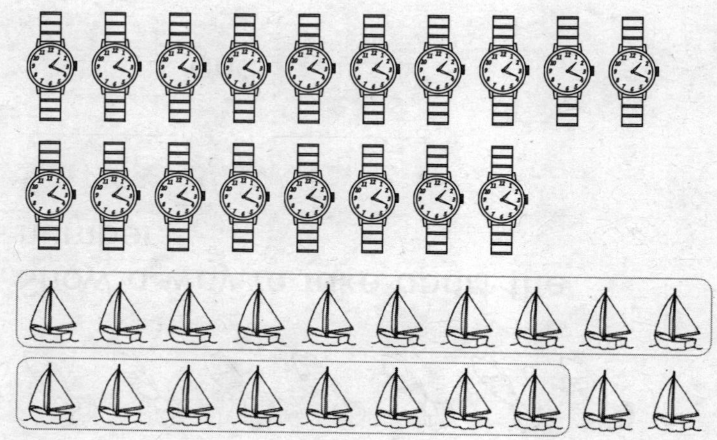

NAME ____________ DATE ____________

Chapter 4 Test

SCORE ____________

1. There are 5 umbrellas. Take apart the number. K.OA.3

1 and 4

Sample answers are shown. Other answers include 2 and 3, 3 and 2, 4 and 1, 0 and 5, 5 and 0.

2. There are 4 flags. Show a way to make 4. K.OA.1

2 and 2

Sample answers are shown. Other answers include 1 and 3, 3 and 1, 4 and 0, 0 and 4.

3. Count the ties. Write the number above the ties. K.OA.3

7

Show a way to take apart the number.

Sample answer:

3 and 4

4. Show a way to make 6 eggs. K.OA.1

4 and 2

5. There are 8 squares. Break apart a group of 5. Write the number in the other group. K.OA.3

8

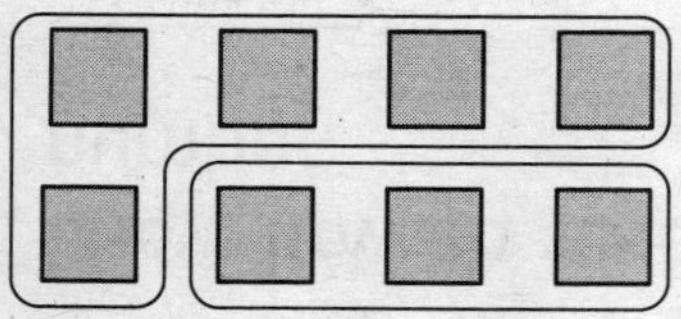

5 and 3

6. Show a way to make 9 by circling screws. Write the numbers. K.OA.1

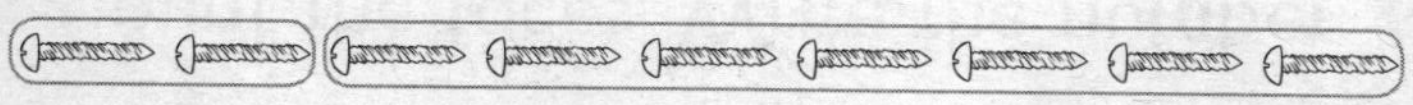

Sample answer:

2 and 7

7. Take apart 10. Write the numbers that make 10. K.OA.3

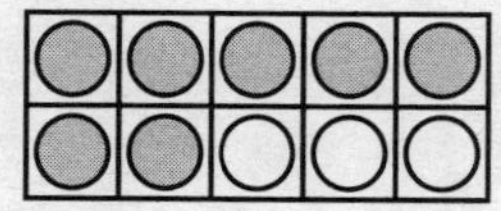

7 and 3

8. Write the number that makes ten. K.OA.4

1 and 9

9. There are 6 shells. Circle shells to make 3 shells and 3 shells. K.OA.1

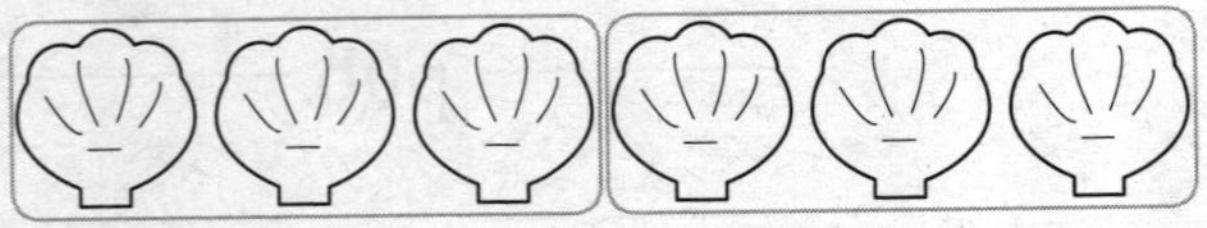

10. Count the faces. Write the number above the faces. K.OA.3

10

Show how to take apart the number.

5   5

11. Match the picture to the statement. K.OA.3

5 and 3 is 8.

3 and 3 is 6.

12. Write the number that makes 9. K.OA.1

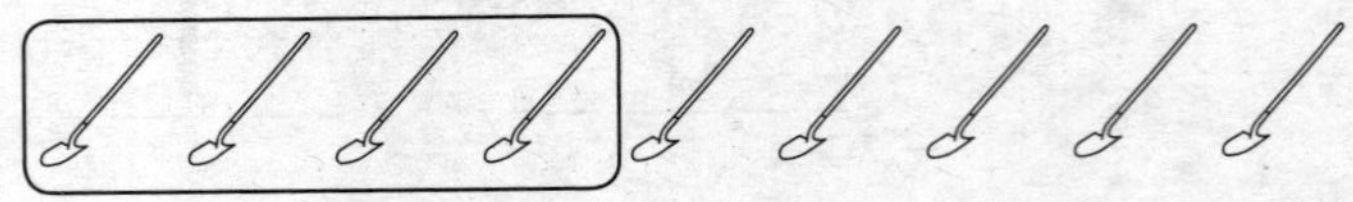

4 and 5

NAME ______________ DATE ______________

SCORE ______________

Chapter 5 Test

1. There are three lizards on the grass. Two more lizards come. How many lizards are there in all? K.OA.1

5

2. Write the number that tells how many there are in all? K.OA.1

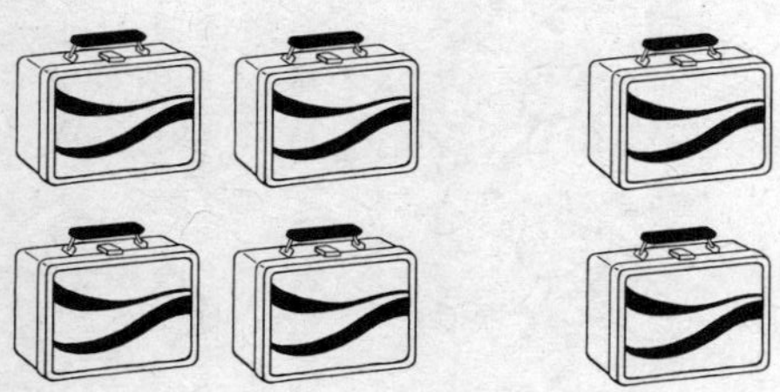

4 + 2 = 6 in all

3. Count the number in each group. Write how many there are in all. K.OA.1

4 + 4 = 8

in all

4. Circle two groups to match the number sentence. K.OA.1

6 + 3 = 9

5. Count the number in each group. Write the numbers. Then write the number in all. K.OA.1

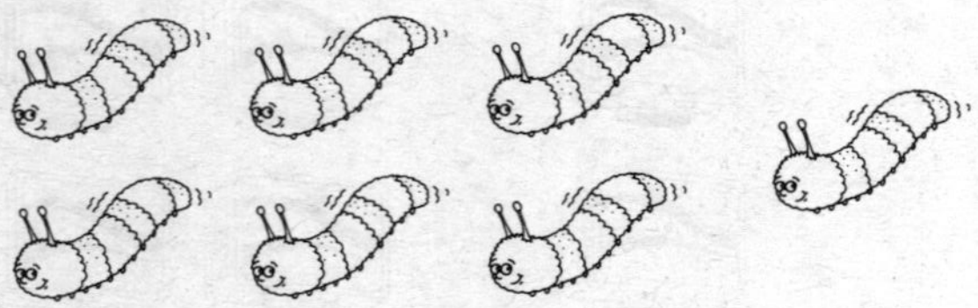

6 + 1 = 7

6. There are 2 cherries. Mrs. Grady buys 2 more. How many cherries are there in all? Write a number sentence. K.OA.1

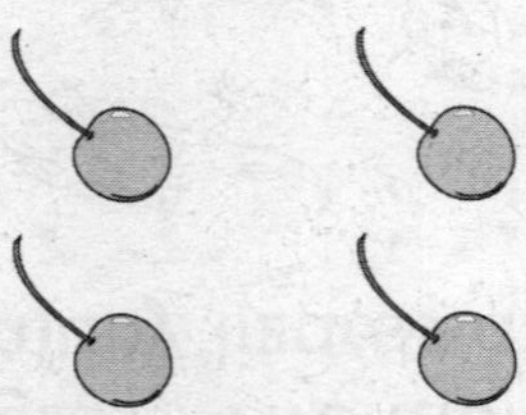

2 + 2 = 4

7. Count the objects and write the numbers. K.OA.4

8 + 2 = 10

8. Huan has 3 crayons. He gets 7 more. How many crayons does he have in all? Circle the correct number. K.OA.4

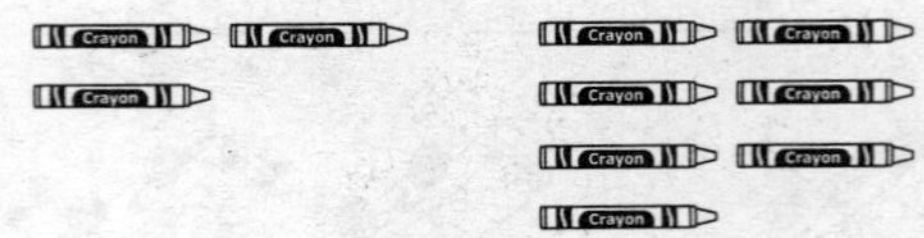

9 (10) 11

Copyright © McGraw-Hill Education. Permission is granted to reproduce for classroom use.

9. Count the number in each group. Write how many there are in all. K.OA.1

3 + 3 = 6 in all

10. Circle the plus sign. K.OA.1

+ − × =

11. Circle the picture that shows 6 and 4 make 10. K.OA.4

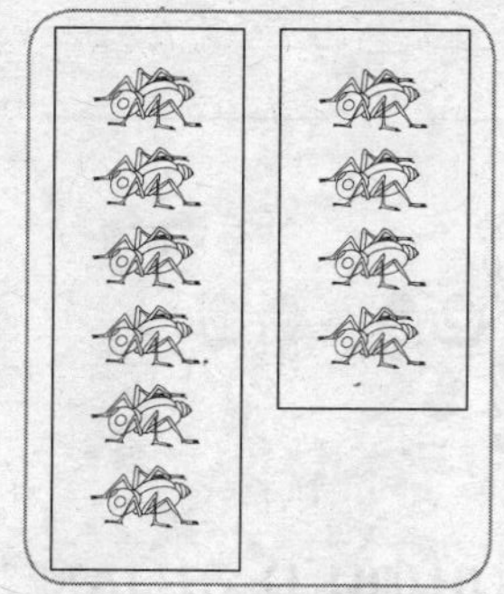

Chapter Tests

NAME ________ DATE ________

SCORE ________

Chapter 6 Test

1. Jaylan saw 5 fish. Two swam away. Cross out two fish. Write how many are left. K.OA.1

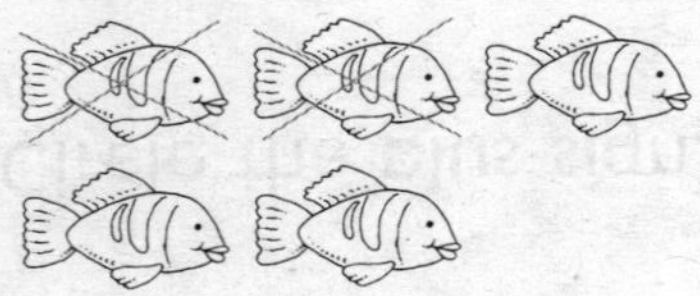

3

2. There were 9 cookies. Jackson ate 6 cookies. Cross out 6 cookies. Write how many cookies are left. K.OA.1

3

3. Count the number of crabs. Count how many are going away. Count how many are left. Write the numbers. K.OA.1

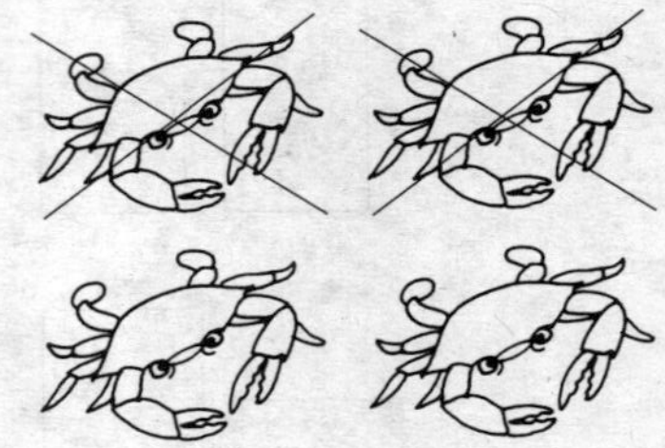

4 – 2 = 2

4. Write a number sentence. K.OA.1

6 – 6 = 0

Copyright © McGraw-Hill Education. Permission is granted to reproduce for classroom use.

5. Count the mice. Seven run away. Cross out seven mice. Write how many are left. K.OA.4

3

6. There are 7 jugs of milk. The students drink 2 of them. How many are left? Write a number sentence. K.OA.1

7 − 2 = 5

7. There are ten airplanes. Five airplanes fly away. Cross out five airplanes. Write how many airplanes are left. K.OA.3

5

8. Henry had 3 apples. He ate one apple. How many apples does Henry have left? Circle the number. K.OA.1

1 (2) 3

9. Cross out acorns to show the number sentence. K.OA.1

$6 - 5 = 1$

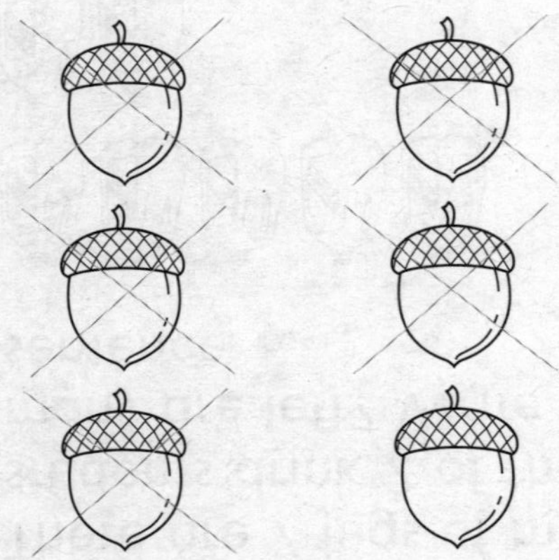

10. Circle the subtraction sign. K.OA.1

+ − = ×

11. Circle the picture that shows $8 - 7 = 1$. K.OA.4

NAME ______________ DATE ______________

Chapter 7 Test

SCORE ______________

1. Count the number in each ten-frame. Write the numbers. Write how many in all. K.NBT.1

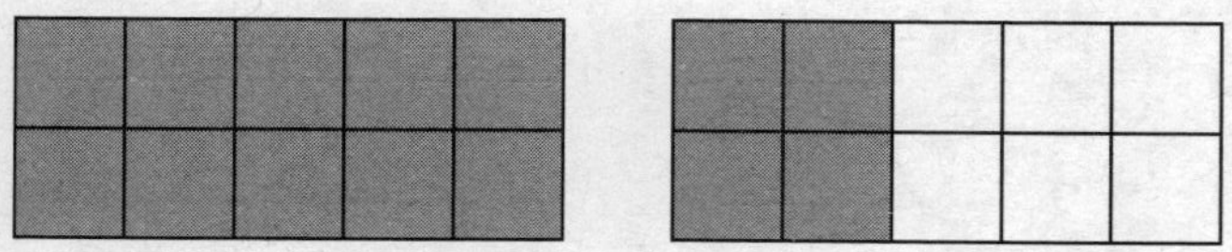

10 and 4 is 14

2. Count the number of each clock. Write each number. K.NBT.1

10   3

3. Shade the ten-frames to show "10 and 7 is 17." K.NBT.1

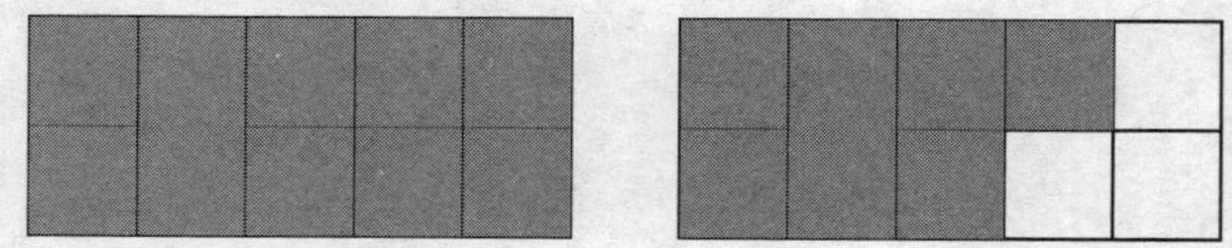

4. There are ten turtles. Six more turtles come. How many turtles are there in all? K.NBT.1

10 and 6 is 16

5. Circle the answer to "ten and five more." K.NBT.1

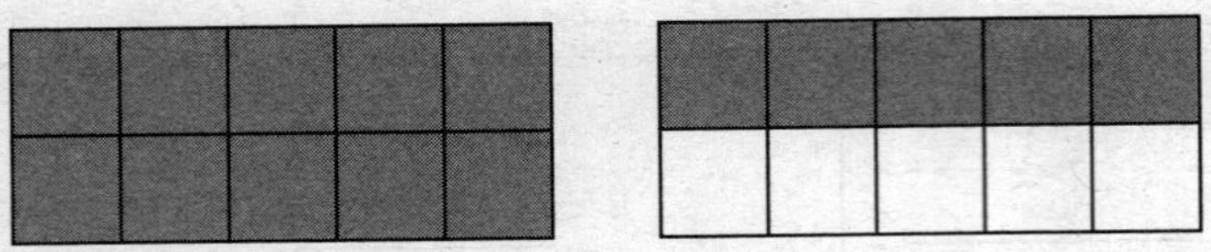

13 14 (15)

6. There were twelve apples in a tree. 2 apples fell off. Cross out two apples. How many apples are left? K.NBT.1

10

7. Count the number of each animal. Write each number. K.NBT.1

10	8

8. There are ten horses. Seven more horses come. How many horses are there in all? K.NBT.1

10 and 7 is 17

9. Which is correct? K.NBT.1

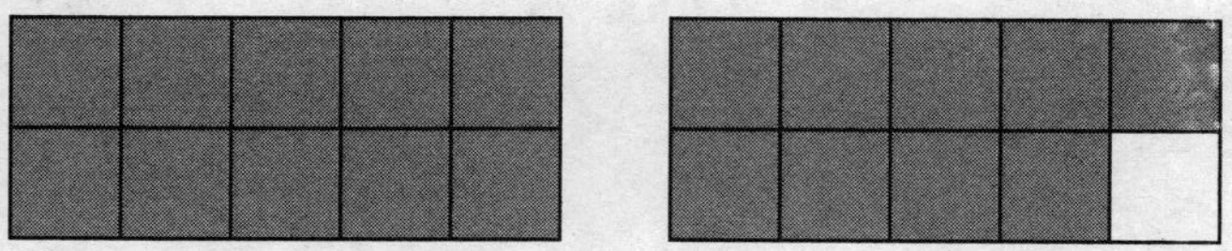

10 and 9 more

10 and 8 more

10 and 7 more

10. There are ten hot air balloons. Ten fly away. Cross out ten balloons. How many are left? K.NBT.1

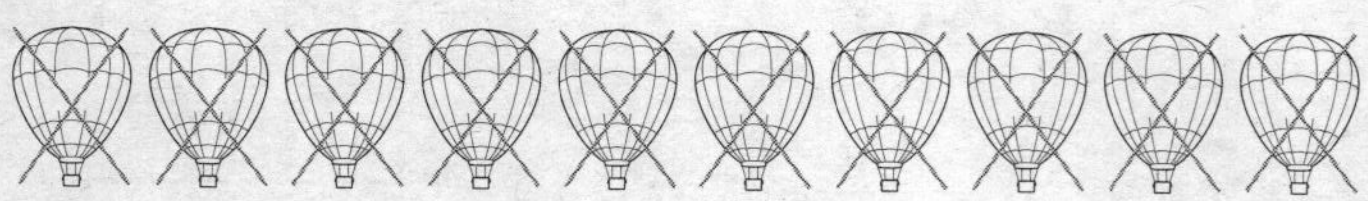

0

11. Circle the picture that shows

10 birds and 4 birds is 14 birds. K.NBT.1

NAME ____________ DATE ____________

SCORE ____________

Chapter 8 Test

1. Compare the objects. Circle the one that is shorter. Underline the one that is longer. K.MD.2

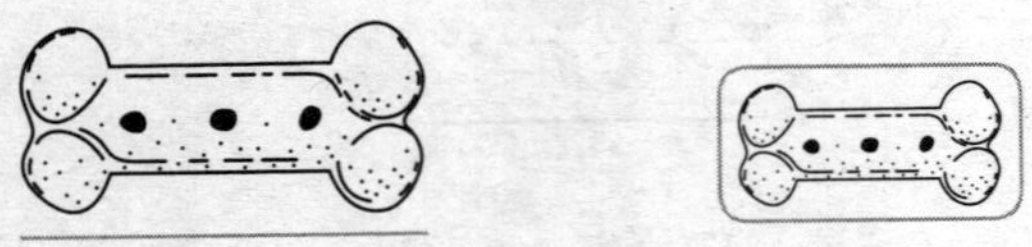

2. Compare the dogs. Circle the taller one. Underline the shorter one. K.MD.2

3. Circle the things that are too heavy to lift. K.MD.2

4. Look at the two objects. Circle the object that is shorter and weighs less than the chair. K.MD.2

5. Match the object to the word. K.MD.2

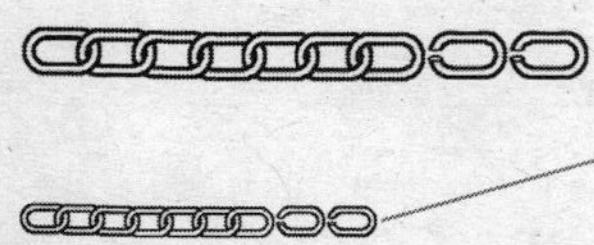

shorter

longer

6. Circle the bowl that holds more. K.MD.2

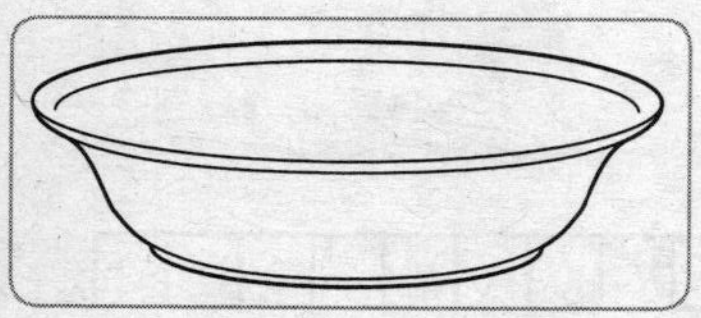

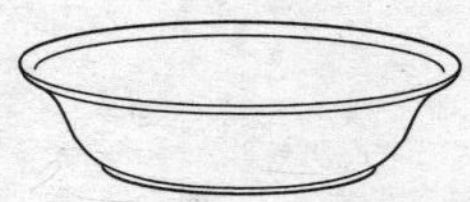

7. Circle the lighter object. K.MD.2

8. How many pennies long is the wagon? K.MD.2

3

9. Underline the containers that hold the same amount. K.MD.2

10. Circle the worm that is 6 blocks long. K.MD.2

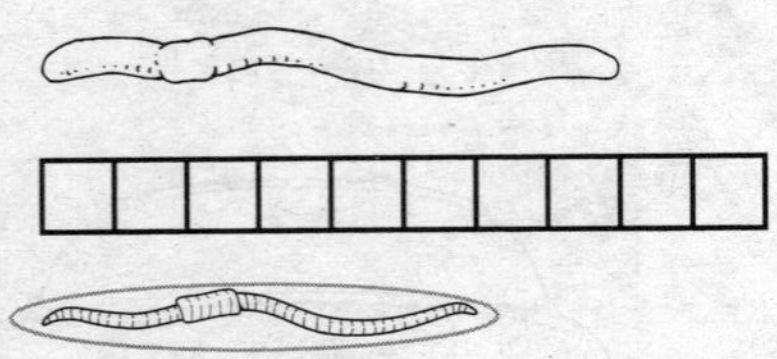

11. Underline the pieces of chalk that are the same length. K.MD.2

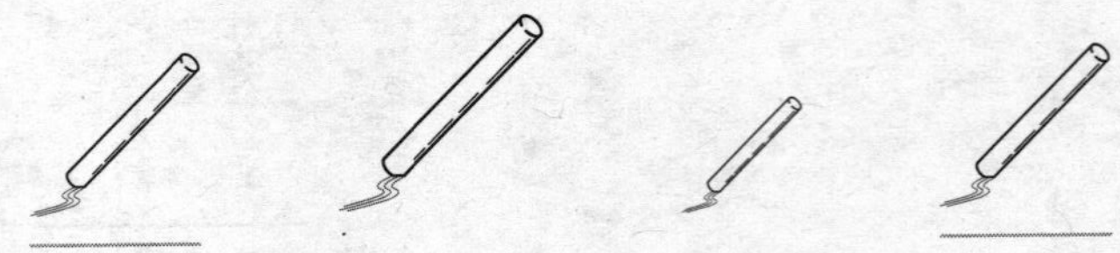

12. How many blocks tall is the cage? K.MD.2

5

Copyright © McGraw-Hill Education. Permission is granted to reproduce for classroom use.

NAME ____________ DATE ____________

Chapter 9 Test

SCORE ____________

1. Circle the ones that are alike. K.MD.3

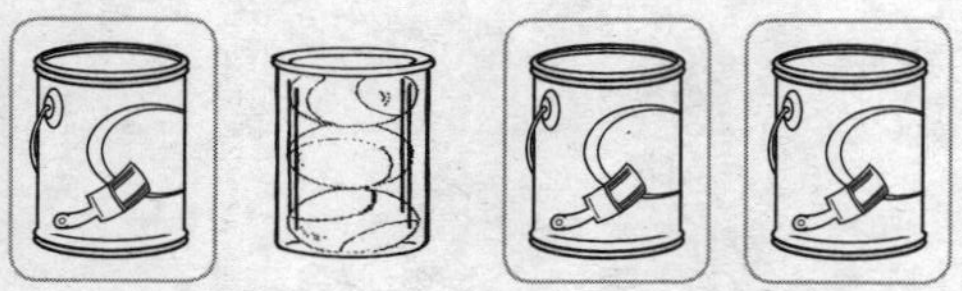

2. Look at the pictures. Circle the pictures that you can eat. K.MD.3

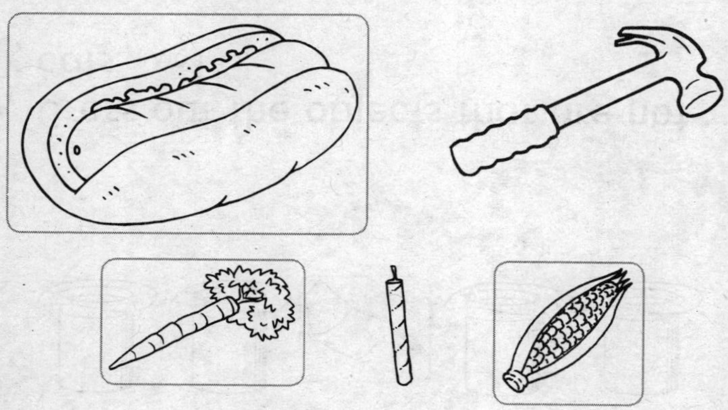

3. Underline the smaller hearts. K.MD.3

4. Circle all the shapes that are alike. K.MD.3

5. Circle the one that is different. K.MD.3

6. Cross out the objects that are not cats. K.MD.3

7. Sort by size. Circle the taller ones. K.MD.3

8. Sort by shape. Cross out the one that is not like the others. K.MD.3

9. Sort by number. Circle the groups that have the same number. K.MD.3

10. Sort by color. Underline all the shapes that have the same color. K.MD.3

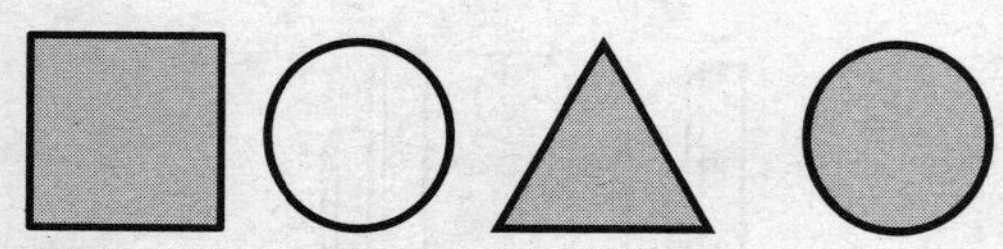

11. Sort by number. Cross out the group that does not have the same number. K.MD.3

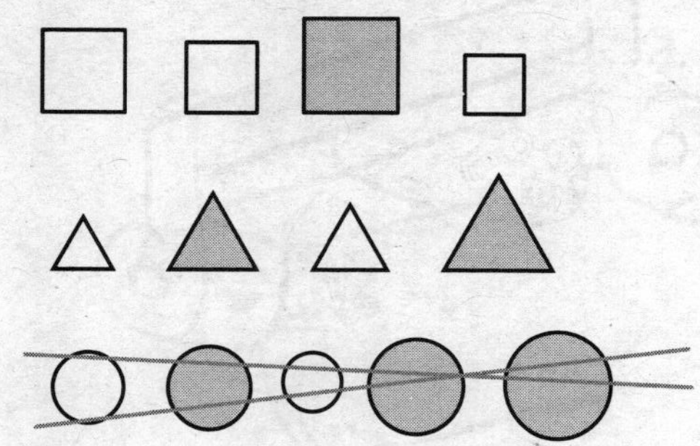

12. Sort by color. Cross out the shape that does not have the same color as the others. K.MD.3

NAME ______________________ DATE ______________

SCORE ______________

Chapter 10 Test

1. Circle the object that is above the table. K.G.I

2. Circle the dog that is in front of the tree. K.G.I

3. Circle the bug that is nearest to the bucket. K.G.I

4. Cross out the objects that are below the couch. K.G.I

5. Cross out the objects that are behind the swing set. K.G.I

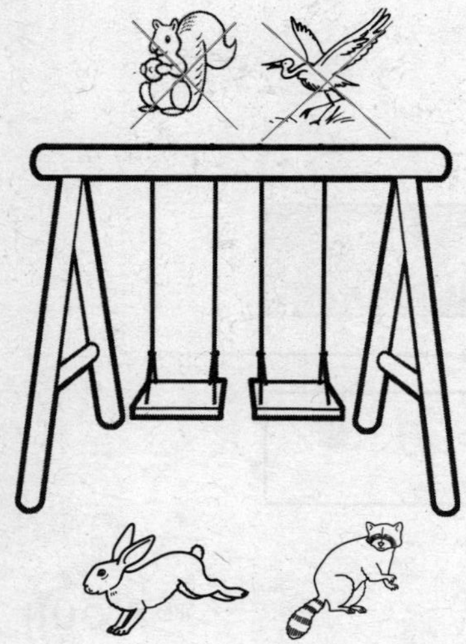

6. Underline the building that is next to the tallest building. K.G.I

7. Draw a circle above the line. K.G.I

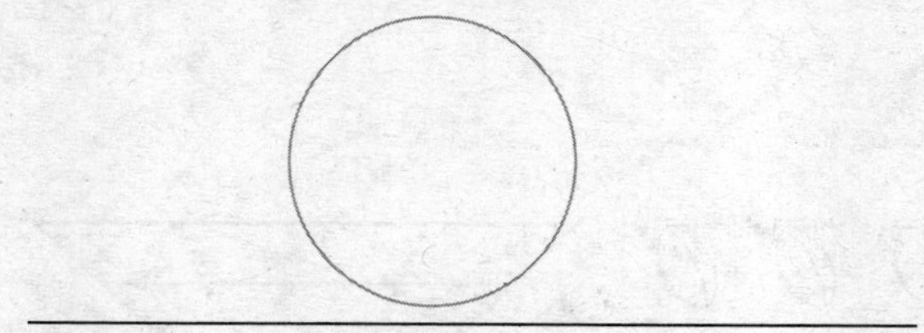

8. Are the shoes in front of or behind the door? Circle the answer. K.G.I

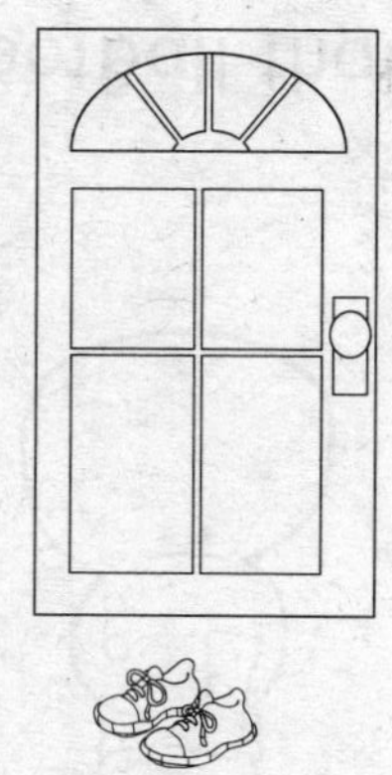

in front of   behind

9. Circle any butterfly that is next to the flower. K.G.I

10. Shade the rectangle below the line. K.G.I

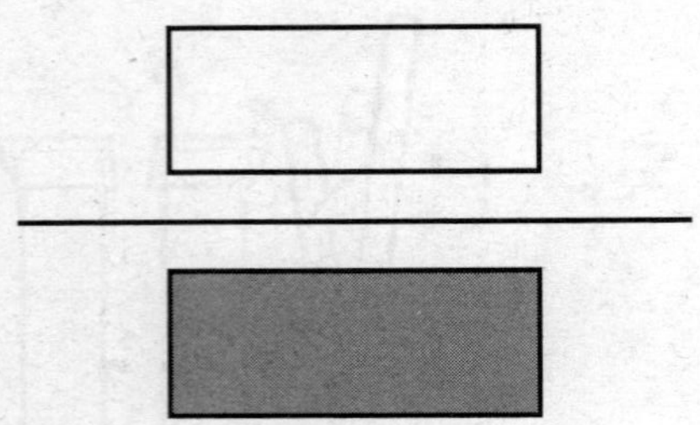

11. Match the two snowmen to the words. K.G.I

12. Circle the football that is nearest to the field. K.G.I

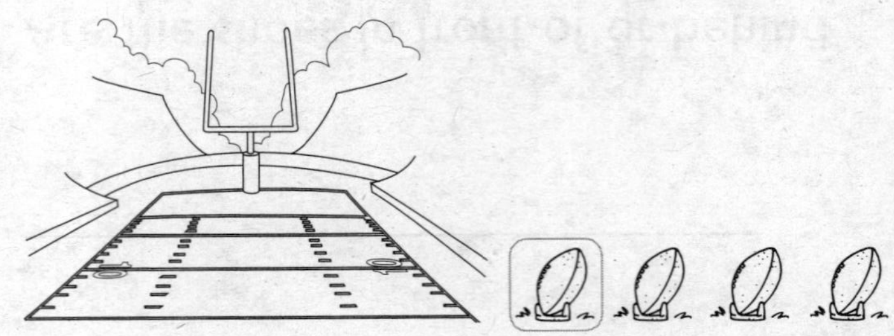

NAME ______________ DATE ______________

Chapter 11 Test

SCORE ______________

1. Shade the squares. K.G.2

2. Underline the object that has a hexagon shape. K.G.2

3. Draw the shape that comes next. K.G.4

4. Circle the shapes that are used to make the design. K.G.5

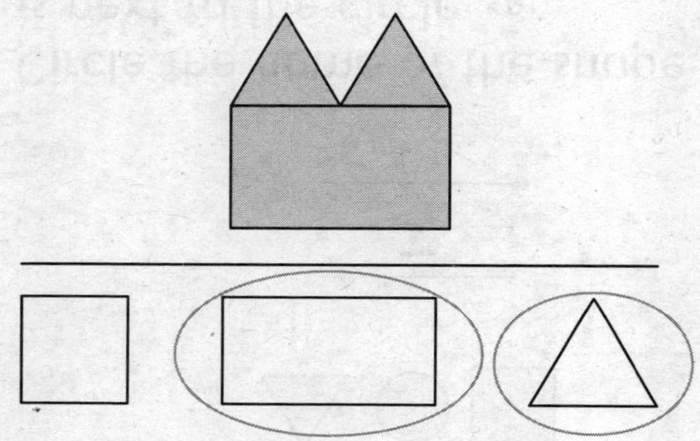

5. Cross out the pictures that do not have circles. K.G.2

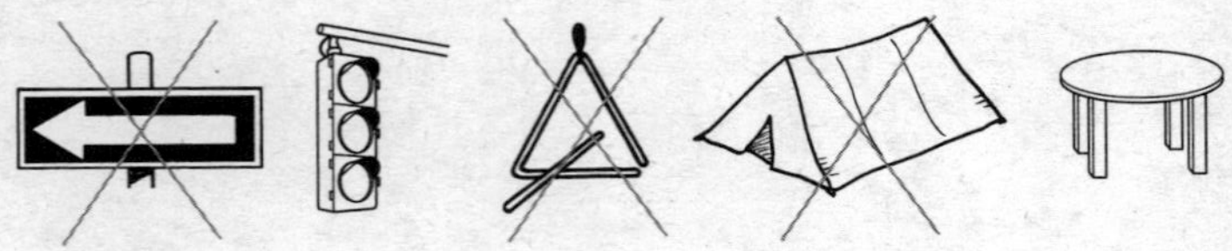

6. Underline the shape that is most like a rectangle. K.G.2

7. Count the number of circles. K.G.4

3

8. Circle the name of the shape that is next to the circle. K.G.1

triangle hexagon square

9. Draw the shape that comes next. K.G.4

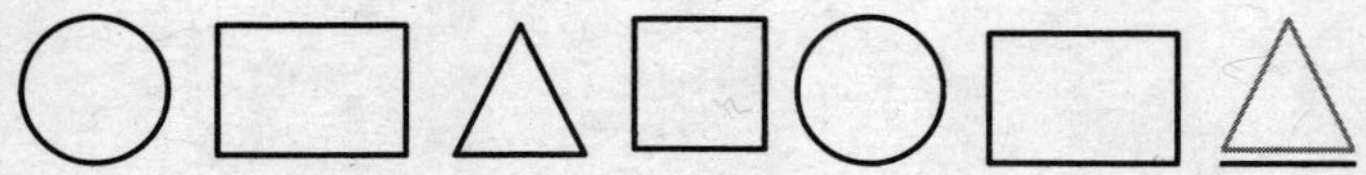

10. Cross out the shapes that are not used to make the design. K.G.5

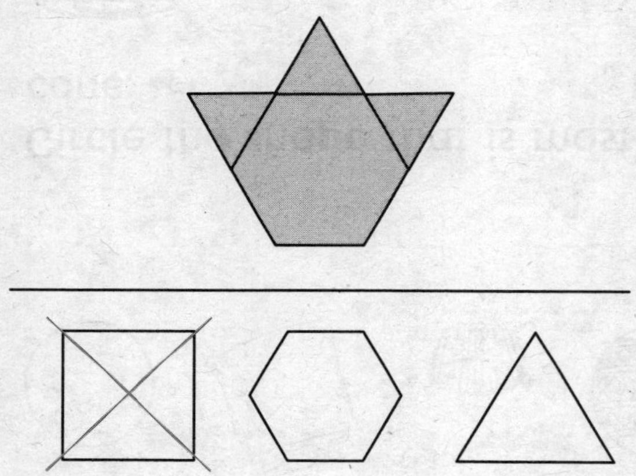

11. Cross out the object that does not have the same shape as the others. K.G.1

12. Circle the shape that is most like the shape of the notes. K.G.1

NAME DATE

Chapter 12 Test

SCORE

1. Put an X on the shapes that are not spheres. K.G.2

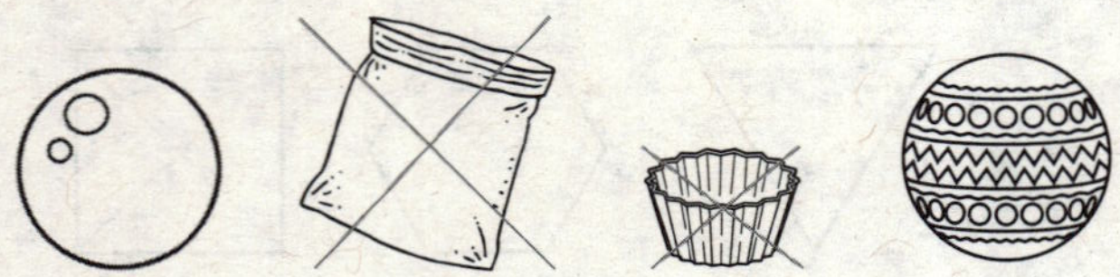

2. Circle the shape that is most like a cone. K.G.1

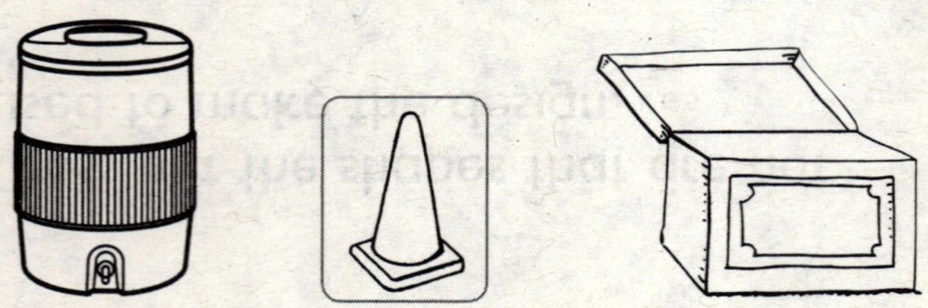

3. Underline the shapes that can slide. K.G.4

4. Circle the shape that best matches the tomato. K.G.2

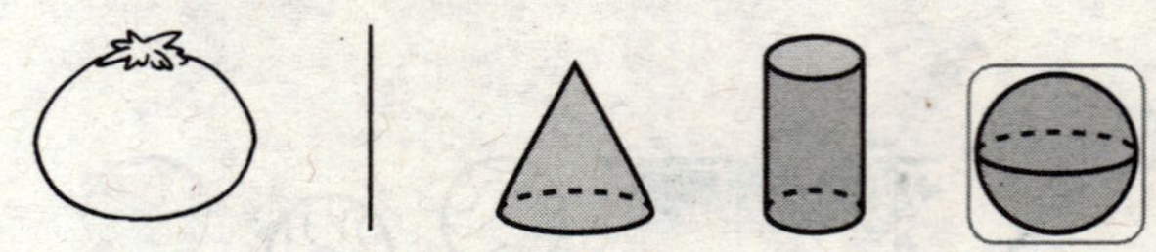

5. Circle the three-dimensional shapes. K.G.1, K.G.2

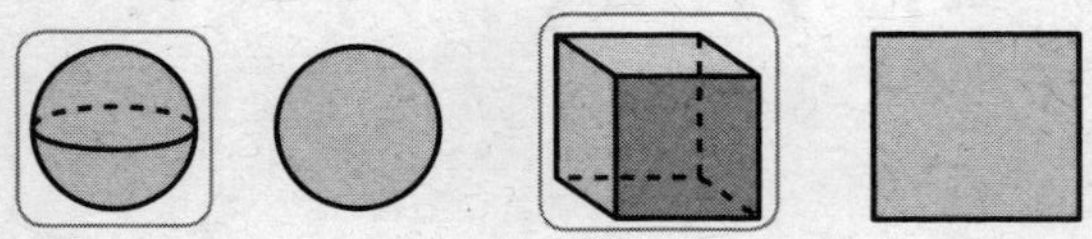

6. Underline the three-dimensional shape that is made from all squares. K.G.2

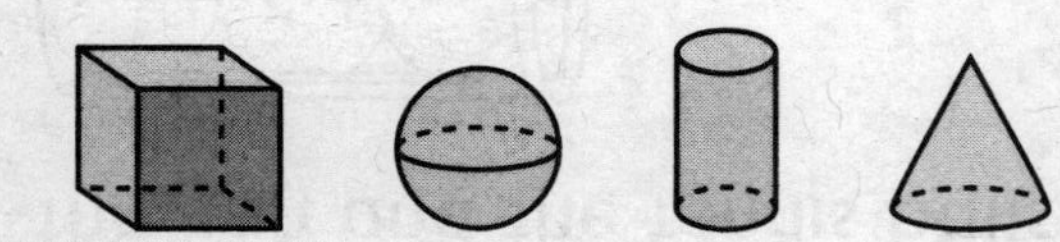

7. Cross out the shape that will not roll. K.G.4

8. Circle the shapes can be stacked. K.G.4

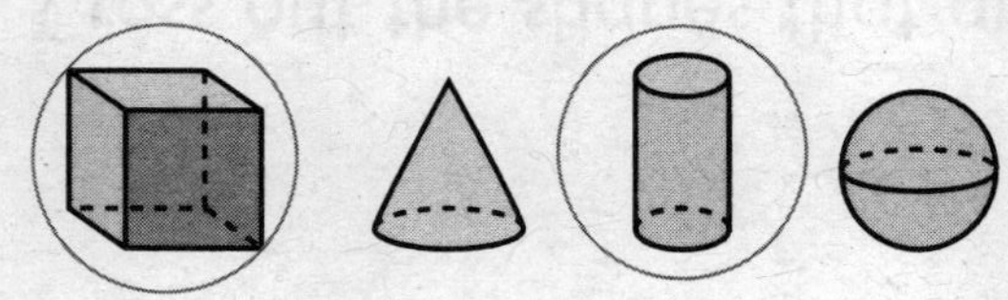

9. Cross out the shape that is different. K.G.2

10. Underline the shapes that make the can and the tennis balls. K.G.2

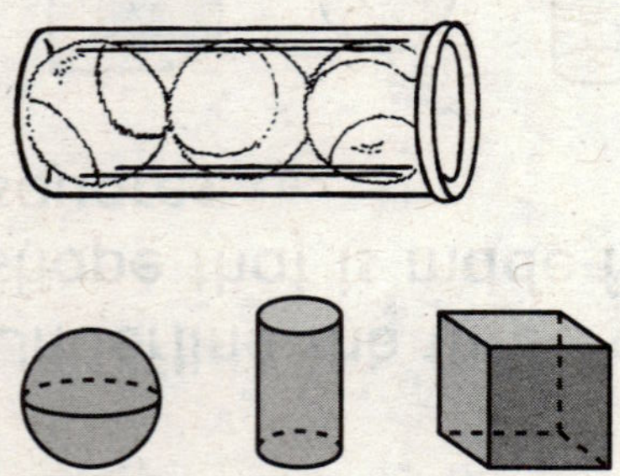

11. Count the spheres in the basket. Write the number. K.G.2

20

12. Cross out the shapes that are not in the picture. K.G.2

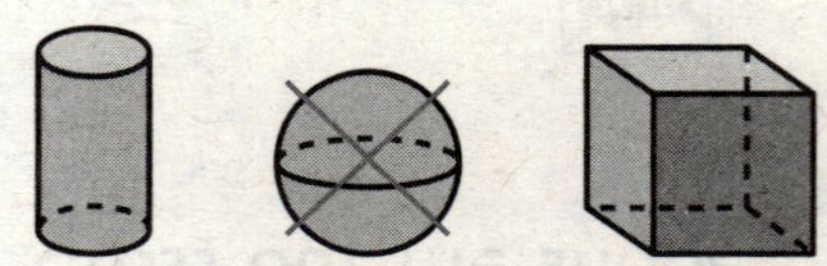

Page 125 • Making Dinner

Task Scenario		
Students will count objects using numbers 1 through 5, write numbers, identify 0, and compare sizes of groups.		
CCSS Content Standard(s)		K.CC.1, K.CC.3, K.CC.4, K.CC.5, K.CC.6
Mathematical Practices		MP1, MP2, MP4
Depth of Knowledge		DOK2, DOK3, DOK4
Part	**Maximum Points**	**Scoring Rubric**
A	1	Full Credit: The student circles the bowl with 0 cherries. No credit will be given for an incorrect answer.
B	2	Full Credit: The student correctly identifies that there are 4 pieces and writes the number clearly. Partial Credit (1 point) will be given for identifying that there are 4 pieces but writing 4 incorrectly. No credit will be given for an incorrect answer.
C	1	Full Credit: 5 No credit will be given for an incorrect answer.
D	2	Full Credit: Student circles the entire group of 5 tomatoes. Partial Credit (1 point) will be given for students who circle part of the group of 5, but not all of it. No credit will be given for an incorrect answer.
E	2	Full Credit: Student circles the entire group of 4 cupcakes. Partial Credit (1 point) will be given for students who circle part of the group of 4, but not all of it. No credit will be given for an incorrect answer.
TOTAL	8	

NAME ____________ DATE ____________

SCORE ____________

Performance Task

Making Dinner

Mrs. Jackson is making dinner.

Write your answers on another piece of paper. Show all your work to receive full credit.

Part A

K.CC.6

Circle the bowl that has zero cherries.

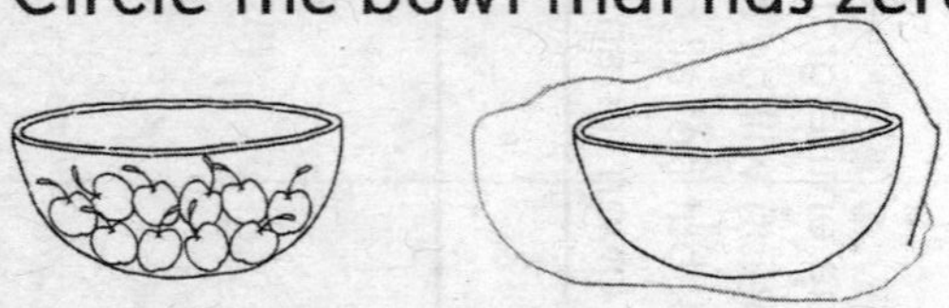

Part B

K.CC.3, K.CC.5

Count the number of broccoli pieces.
Write the number.

Performance Task *(continued)*

Part C

K.CC.3, K.CC.5

Count the number of forks. Circle the number that is one more than that.

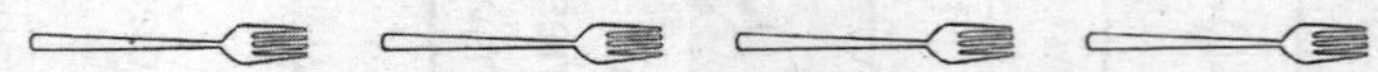

3 4 5

Part D

K.CC.1, K.CC.4

Circle the group with 5 tomatoes.

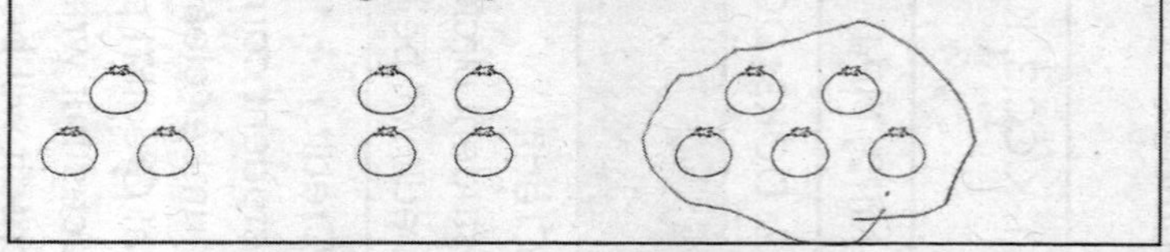

Part E

K.CC.1, K.CC.4

Circle the group with more cupcakes.

Copyright © McGraw-Hill Education. Permission is granted to reproduce for classroom use.

NAME ____________ DATE ____________

SCORE ____________

Performance Task

Making Dinner

Mrs. Jackson is making dinner.

Write your answers on another piece of paper. Show all your work to receive full credit.

Part A

K.CC.6

Circle the bowl that has zero cherries.

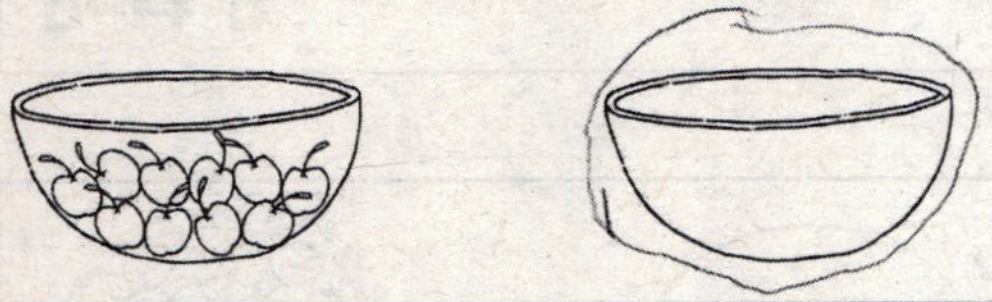

Part B

K.CC.3, K.CC.5

Count the number of broccoli pieces. Write the number.

4

Performance Task *(continued)*

Part C

K.CC.3, K.CC.5

Count the number of forks. Circle the number that is one more than that.

3 4 5

Part D

K.CC.1, K.CC.4

Circle the group with 5 tomatoes.

Part E

K.CC.1, K.CC.4

Circle the group with more cupcakes.

Student Model

NAME ______________ DATE ______________

SCORE ______________

Performance Task

Making Dinner

Mrs. Jackson is making dinner.

Write your answers on another piece of paper. Show all your work to receive full credit.

Part A

K.CC.6

Circle the bowl that has zero cherries.

Part B

K.CC.3, K.CC.5

Count the number of broccoli pieces. Write the number.

3

Performance Task *(continued)*

Part C

K.CC.3, K.CC.5

Count the number of forks. Circle the number that is one more than that.

3 4 5

Part D

K.CC.1, K.CC.4

Circle the group with 5 tomatoes.

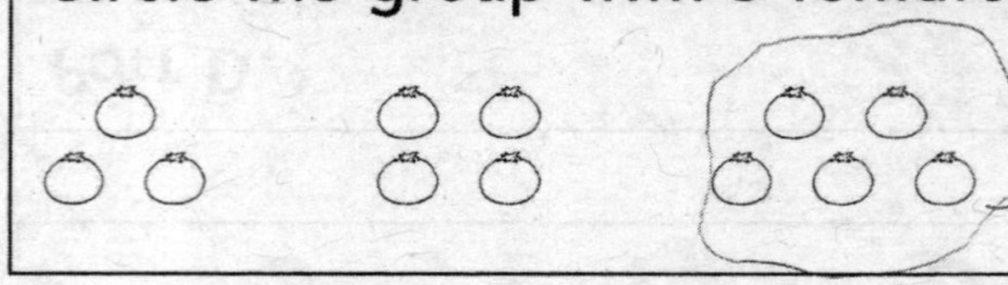

Part E

K.CC.1, K.CC.4

Circle the group with more cupcakes.

NAME ____________ DATE ____________

SCORE ____________

Performance Task

Making Dinner

Mrs. Jackson is making dinner.

Write your answers on another piece of paper. Show all your work to receive full credit.

Part A

K.CC.6

Circle the bowl that has zero cherries.

Part B

K.CC.3, K.CC.5

Count the number of broccoli pieces. Write the number.

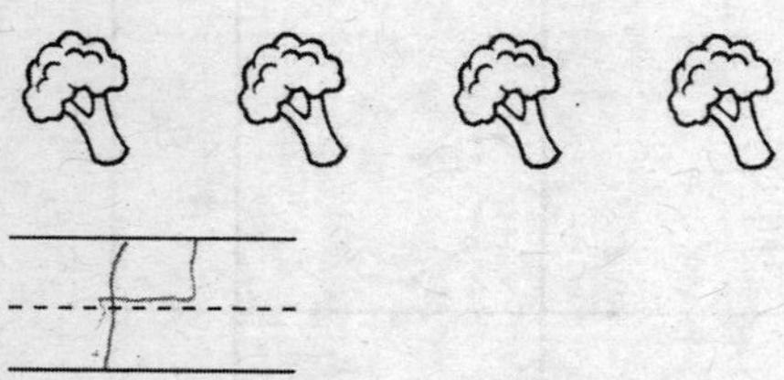

Performance Task *(continued)*

Part C

K.CC.3, K.CC.5

Count the number of forks. Circle the number that is one more than that.

3 4 5

Part D

K.CC.1, K.CC.4

Circle the group with 5 tomatoes.

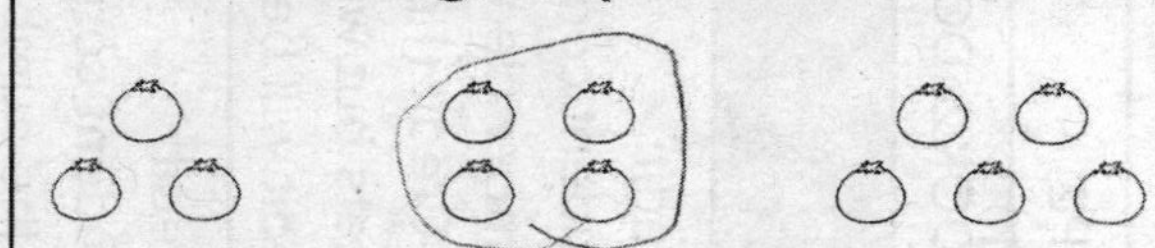

Part E

K.CC.1, K.CC.4

Circle the group with more cupcakes.

Student Model

Page 127 • At the Cat Shelter

Task Scenario Students will count objects, write the associated number, compare numbers using one-to-one correspondence, and draw objects to represent an amount.	
CCSS Content Standard(s)	K.CC.3, K.CC.4, K.CC.5, K.CC.6
Mathematical Practices	MP1, MP2
Depth of Knowledge	DOK1, DOK2, DOK3

Part	Maximum Points	Scoring Rubric
A	2	Full Credit: The student correctly identifies that there are eightcats and writes the number clearly. Partial Credit (1 point) will be given for identifying that there are eight cats but writing 8 incorrectly. No credit will be given for an incorrect answer.
B	2	Full Credit: The student correctly identifies that there are nine bowls and writes the number clearly. Partial Credit (1 point) will be given for identifying that there are nine bowls but writing 9 incorrectly. No credit will be given for an incorrect answer.
C	1	Full Credit: The student correctly matches the items using one-to-one correspondence, connecting one cat to one bowl and leaving one bowl without a line to a cat. No credit will be given for an incorrect answer.
D	2	Full Credit: The student correctly circles all the cats. Partial Credit (1 point) will be given for circling some, but not all of the cats. No credit will be given if no cats are circled or if any bowls are circled.
E	1	Full Credit: The student correctly identifies that eight cats would require eight balls and draws eight balls. No credit will be given for an incorrect answer.
TOTAL	8	

NAME ____________ DATE ____________

SCORE ____________

Performance Task

At the Cat Shelter

You are helping at a cat shelter.

Part A

These are the cats that need help.
Write the number of cats.

8

Part B

You have these bowls for cat food.
Write the number of bowls.

9

Performance Task *(continued)*

Part C

Match the cats to the bowls.

Part D

Circle the group that is less than the other group.

Part E

Draw one ball for each cat.

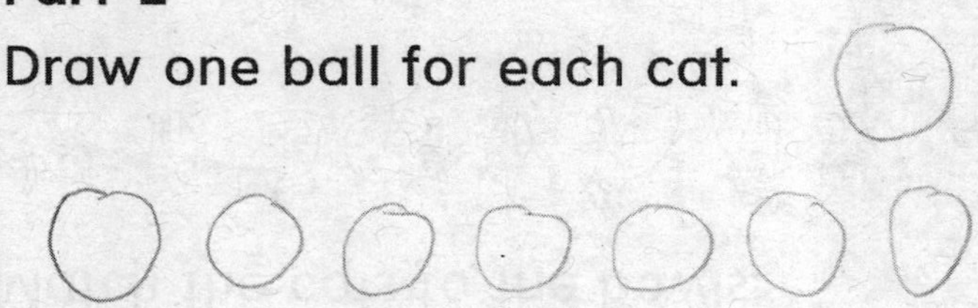

Student Model

NAME ______________________ DATE ______________

SCORE ______________

Performance Task

At the Cat Shelter

You are helping at a cat shelter.

Part A

These are the cats that need help.
Write the number of cats.

______

Part B

You have these bowls for cat food.
Write the number of bowls.

______

Performance Task *(continued)*

Part C

Match the cats to the bowls.

Part D

Circle the group that is less than the other group.

Part E

Draw one ball for each cat.

NAME ______________ DATE ______________

SCORE ______________

Performance Task

At the Cat Shelter

You are helping at a cat shelter.

Part A

These are the cats that need help.
Write the number of cats.

Part B

You have these bowls for cat food.
Write the number of bowls.

Performance Task *(continued)*

Part C

Match the cats to the bowls.

Part D

Circle the group that is less than the other group.

Part E

Draw one ball for each cat.

Student Model

NAME ________________ DATE ________________

SCORE ________________

Performance Task

At the Cat Shelter

You are helping at a cat shelter.

Part A

These are the cats that need help.
Write the number of cats.

9

Part B

You have these bowls for cat food.
Write the number of bowls.

10

Performance Task *(continued)*

Part C

Match the cats to the bowls.

Part D

Circle the group that is less than the other group.

Part E

Draw one ball for each cat.

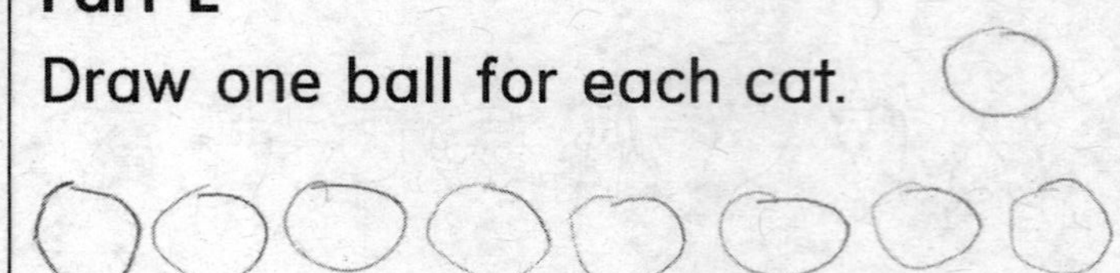

Page 129 • At the Hat Shop

Task Scenario Students will count objects, compare numbers from 10 to 20, and count by 10s starting at 10.	
CCSS Content Standard(s)	K.CC.3, K.CC.5, K.CC.6
Mathematical Practices	MP1, MP2, MP4
Depth of Knowledge	DOK1, DOK2, DOK3

Part	Maximum Points	Scoring Rubric
A	2	Full Credit: The student correctly identifies that there are 15 hats and writes the number clearly. Partial Credit (1 point) will be given for identifying that there are 15 hats but writing 15 incorrectly. No credit will be given for an incorrect answer.
B	2	Full Credit: The student correctly identifies that there are 18 boxes and writes the number clearly. Partial Credit (1 point) will be given for identifying that there are 18 boxes but writing 18 incorrectly. No credit will be given for an incorrect answer.
C	2	Full Credit: The student correctly matches one hat to one box and correctly circles the entire group of boxes. Partial Credit (1 point) will be given for a correct matching **OR** a correct circling of all the boxes. No credit will be given for two incorrect responses.
D	1	Full Credit: The student correctly shades all 20 squares. No credit will be given for incorrect shading.
E	2	Full Credit: 20, 30 Partial Credit (1 point) will be given for 20 **OR** 30. No credit will be given for two incorrect responses.
TOTAL	9	

NAME ________ DATE ________

SCORE ________

Performance Task

At the Hat Shop

You are putting hats in boxes at a hat shop.

Write your answers on another piece of paper. Show all your work to receive full credit.

Part A

K.CC.3, K.CC.5

These are the hats that need boxed. Write the number of hats.

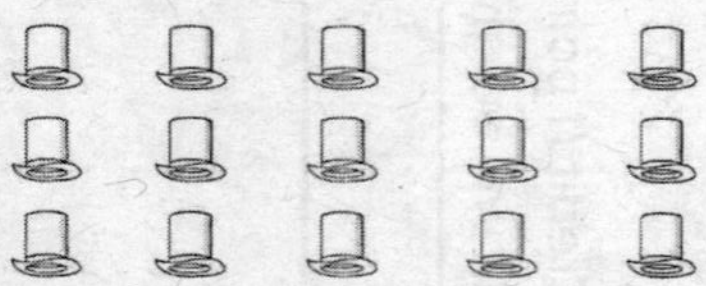

15

Part B

K.CC.3, K.CC.5

These are the boxes. Write the number of boxes.

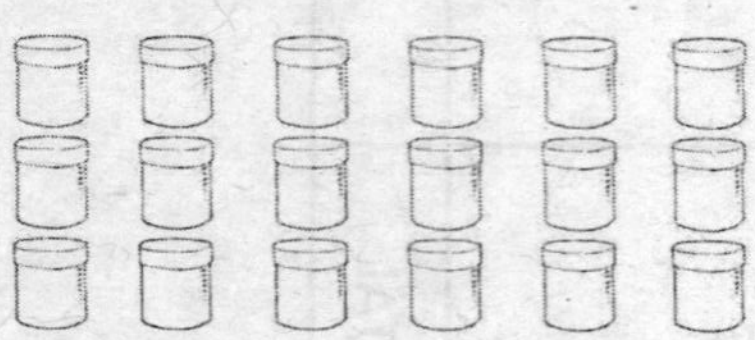

18

Performance Task *(continued)*

Part C

K.CC.3, K.CC.6

Match the hats to the boxes. Circle the group that has more.

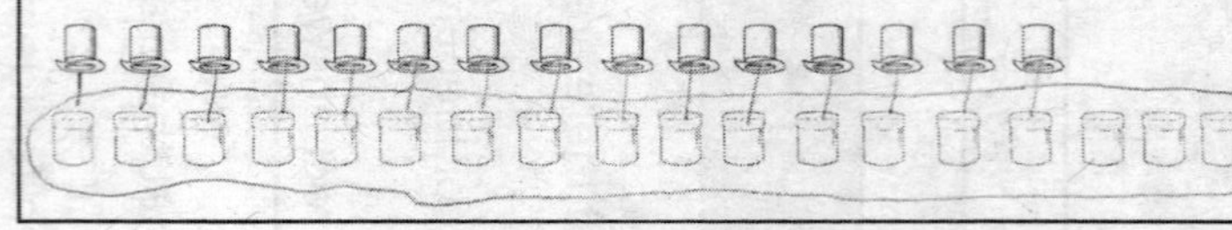

Part D

K.CC.3, K.CC.5

There are 20 price tags. Color the ten-frames to show the number of tags.

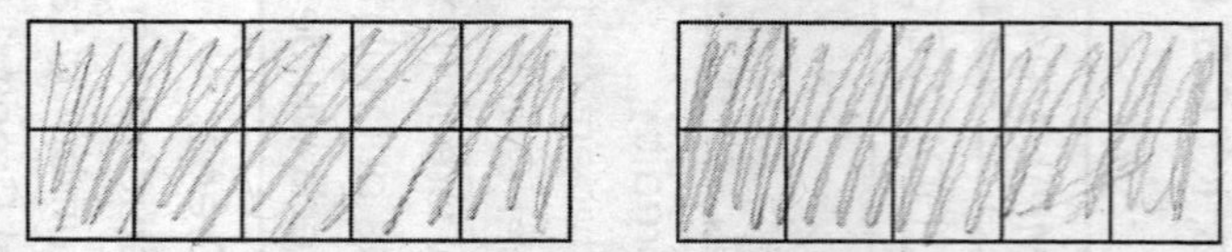

Part E

K.CC.3, K.CC.5

Count the price tags by 10.

10 20 30

NAME ______________ DATE ______________

SCORE ______________

Performance Task

At the Hat Shop

You are putting hats in boxes at a hat shop.

Write your answers on another piece of paper. Show all your work to receive full credit.

Part A

K.CC.3, K.CC.5

These are the hats that need boxed. Write the number of hats.

15

Part B

K.CC.3, K.CC.5

These are the boxes. Write the number of boxes.

18

Performance Task *(continued)*

Part C

K.CC.3, K.CC.6

Match the hats to the boxes. Circle the group that has more.

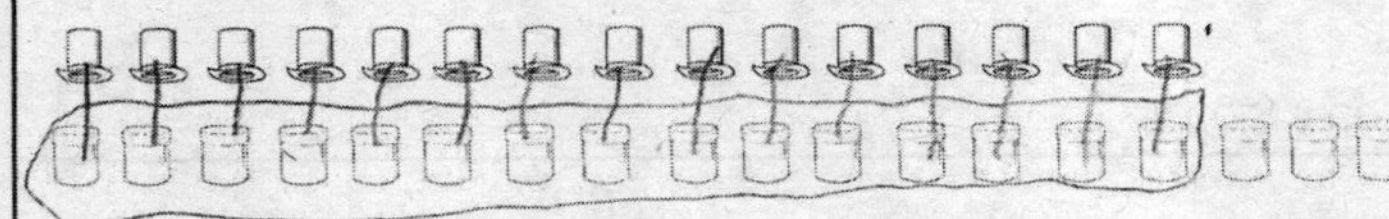

Part D

K.CC.3, K.CC.5

There are 20 price tags. Color the ten-frames to show the number of tags.

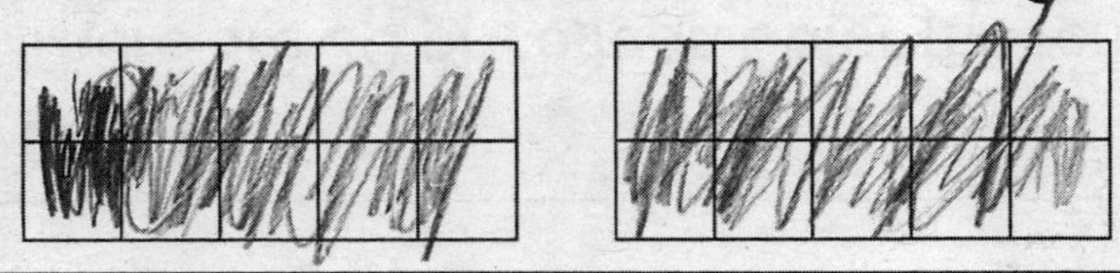

Part E

K.CC.3, K.CC.5

Count the price tags by 10.

10 20 30

Student Model

NAME ______________________ DATE ____________

SCORE ____________

Performance Task

At the Hat Shop

You are putting hats in boxes at a hat shop.

Write your answers on another piece of paper. Show all your work to receive full credit.

Part A

K.CC.3, K.CC.5

These are the hats that need boxed. Write the number of hats.

15

Part B

K.CC.3, K.CC.5

These are the boxes. Write the number of boxes.

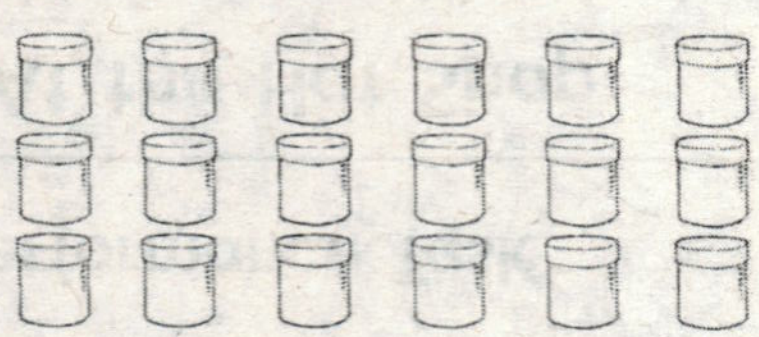

17

Performance Task *(continued)*

Part C

K.CC.3, K.CC.6

Match the hats to the boxes. Circle the group that has more.

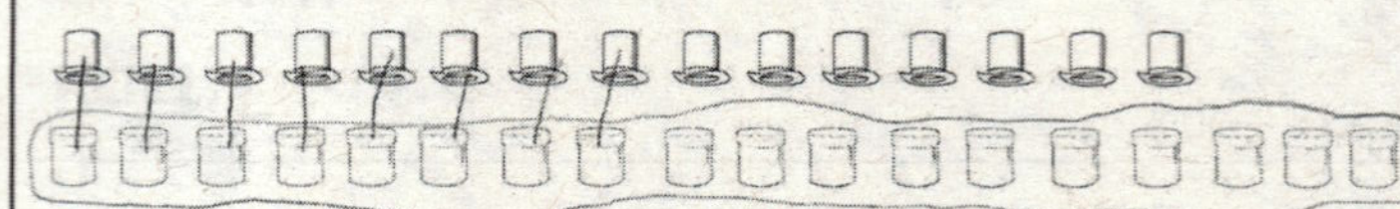

Part D

K.CC.3, K.CC.5

There are 20 price tags. Color the ten-frames to show the number of tags.

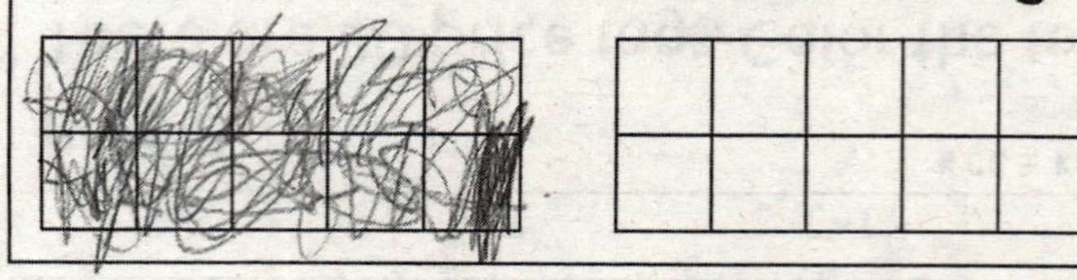

Part E

K.CC.3, K.CC.5

Count the price tags by 10.

10 20 50

NAME ______________ DATE ______________

SCORE ______________

Performance Task

At the Hat Shop

You are putting hats in boxes at a hat shop.

Write your answers on another piece of paper. Show all your work to receive full credit.

Part A

K.CC.3, K.CC.5

These are the hats that need boxed. Write the number of hats.

11

Part B

K.CC.3, K.CC.5

These are the boxes. Write the number of boxes.

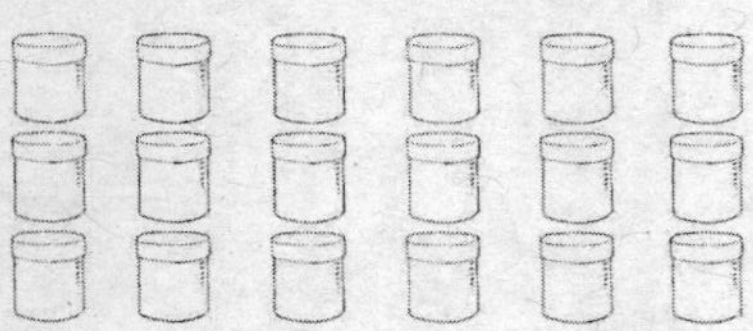

8

Performance Task *(continued)*

Part C

K.CC.3, K.CC.6

Match the hats to the boxes. Circle the group that has more.

Part D

K.CC.3, K.CC.5

There are 20 price tags. Color the ten-frames to show the number of tags.

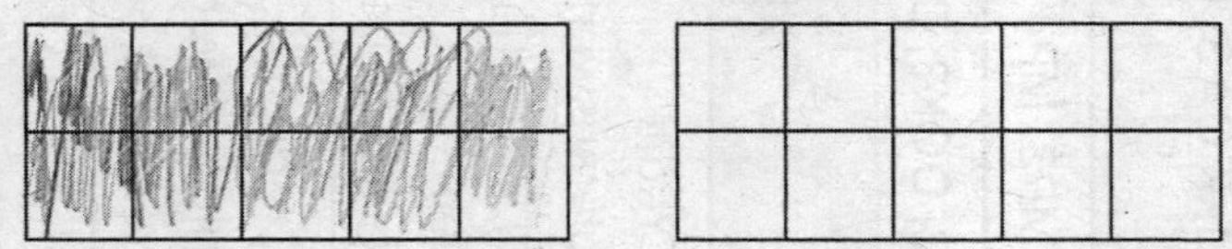

Part E

K.CC.3, K.CC.5

Count the price tags by 10.

10 10 20

Student Model

Page 131 • Raising Rabbits

Task Scenario Students will use numbers 1 through 10, combine them, and break them apart to count objects.	
CCSS Content Standard(s)	K.OA.1, K.OA.3, K.OA.4
Mathematical Practices	MP1, MP2, MP4
Depth of Knowledge	DOK2, DOK3, DOK4

Part	Maximum Points	Scoring Rubric
A	2	Full Credit: Sample answer: 4 and 2, 3 and 3. Partial Credit (1 point) will be given for having only one correct answer. No credit will be given for two incorrect responses.
B	1	Full Credit: Sample answer: 2 and 6 No credit will be given for an incorrect answer.
C	2	Full Credit: Sample answer: 3 and 4. Student circles a group of 3 and a separate circle around the remaining 4. Partial Credit (1 point) will be given for a correct answer **OR** a correct circling. No credit will be given for an incorrect answer AND an incorrect circling.
D	1	Full Credit: 3 No credit will be given for an incorrect answer.
TOTAL	6	

NAME ______________ DATE ______________

SCORE ______________

Performance Task

Raising Rabbits

You are raising rabbits.

Write your answers on another piece of paper. Show all your work to receive full credit.

Part A

K.OA.1

There are six rabbits. Show two different ways to make 6.

4 and 2

3 and 3

Part B

K.OA.1

Show a way to make 8 cabbages. Write the numbers.

2 and 6

Performance Task *(continued)*

Part C

K.OA.3

There are 7 carrots. Break apart the carrots into two groups. Circle your groups. Write the numbers.

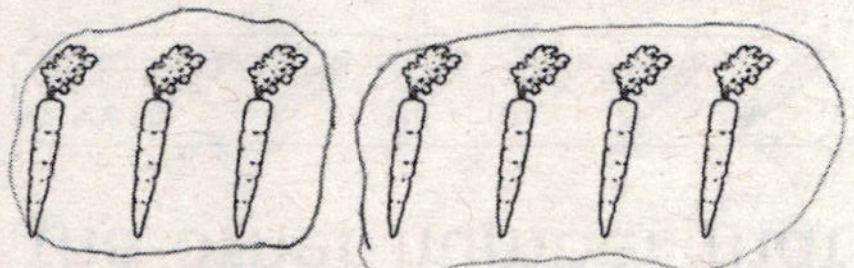

3 and 4

Part D

K.OA.4

There are ten ears of corn. Write the other number that makes ten.

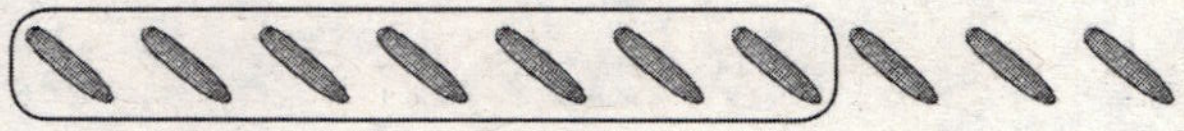

7 and 3

Student Model

NAME ____________ DATE ____________

SCORE ____________

Performance Task

Raising Rabbits

You are raising rabbits.

Write your answers on another piece of paper. Show all your work to receive full credit.

Part A

K.OA.1

There are six rabbits. Show two different ways to make 6.

5 and 1

0 and 6

Part B

K.OA.1

Show a way to make 8 cabbages. Write the numbers.

1 and 7

Performance Task *(continued)*

Part C

K.OA.3

There are 7 carrots. Break apart the carrots into two groups. Circle your groups. Write the numbers.

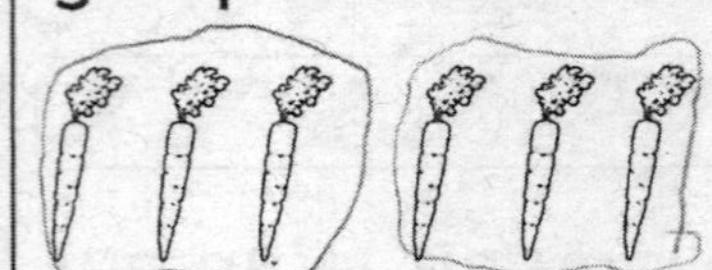

3 and 4

Part D

K.OA.4

There are ten ears of corn. Write the other number that makes ten.

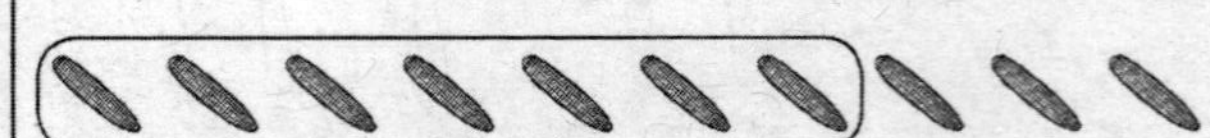

7 and 5

NAME ______________ DATE ______________

SCORE ______________

Performance Task

Raising Rabbits

You are raising rabbits.

Write your answers on another piece of paper. Show all your work to receive full credit.

Part A

K.OA.1

There are six rabbits. Show two different ways to make 6.

4 and 1

3 and 3

Part B

K.OA.1

Show a way to make 8 cabbages. Write the numbers.

2 and 5

Performance Task *(continued)*

Part C

K.OA.3

There are 7 carrots. Break apart the carrots into two groups. Circle your groups. Write the numbers.

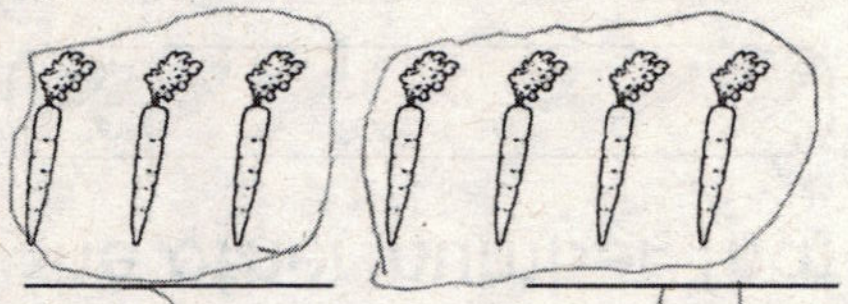

2 and 4

Part D

K.OA.4

There are ten ears of corn. Write the other number that makes ten.

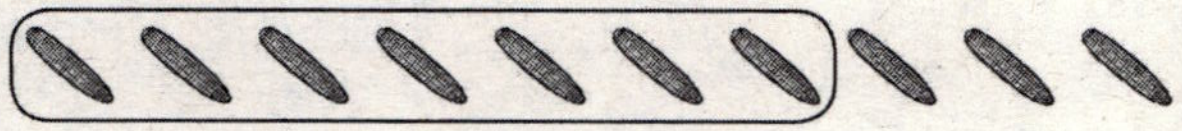

7 and 4

Student Model

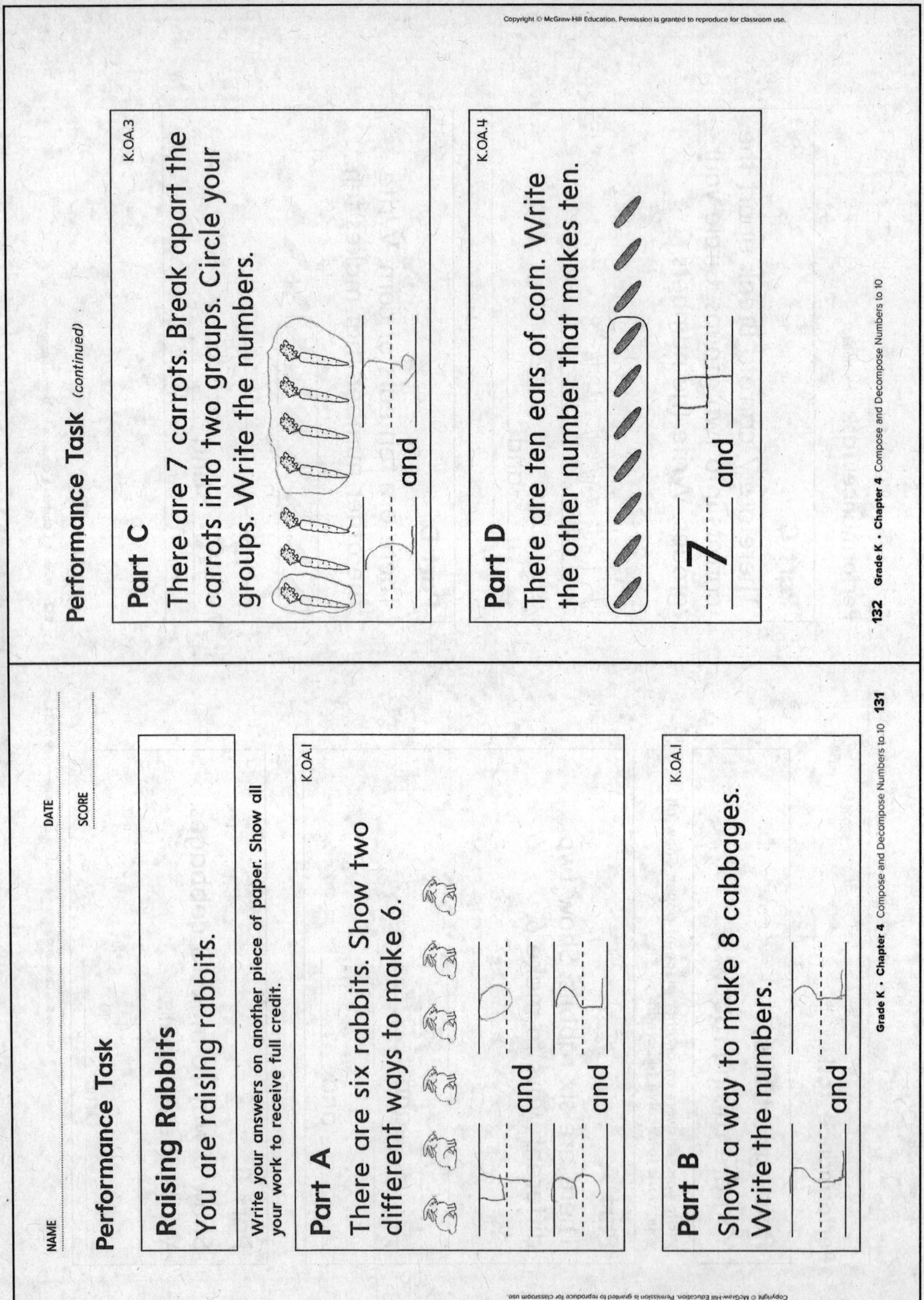

NAME ____ DATE ____

SCORE ____

Performance Task

Raising Rabbits

You are raising rabbits.

Write your answers on another piece of paper. Show all your work to receive full credit.

Part A K.OA.1

There are six rabbits. Show two different ways to make 6.

____ and ____

____ and ____

Part B K.OA.1

Show a way to make 8 cabbages. Write the numbers.

____ and ____

Performance Task *(continued)*

Part C K.OA.3

There are 7 carrots. Break apart the carrots into two groups. Circle your groups. Write the numbers.

____ and ____

Part D K.OA.4

There are ten ears of corn. Write the other number that makes ten.

7 and ____

Page 133 • At the Sports Store

Task Scenario Students will use addition and numbers from 0 to 9 to combine groups of balls.		
CCSS Content Standard(s)		K.OA.1, K.CC.3
Mathematical Practices		MP1, MP2, MP3, MP4
Depth of Knowledge		DOK2, DOK3
Part	**Maximum Points**	**Scoring Rubric**
A	2	Full Credit: $5 + 3 = 8$ Partial Credit (1 point) will be given for the correct parts (5 and 3) **OR** the correct total (8). No credit will be given for incorrect parts AND an incorrect total.
B	2	Full Credit: A circle around 8 baseballs and a separate circle around the remaining 1 baseball. Partial Credit (1 point) will be given for a circle around 8 **OR** a circle around 1. No credit will be given for two incorrect circles.
C	2	Full Credit: $4 + 2 = 6$ Partial Credit (1 point) will be given for the correct parts (4 and 2) **OR** the correct total (6). No credit will be given for incorrect parts AND an incorrect total.
D	1	Full Credit: 12 No credit will be given for an incorrect answer.
TOTAL	7	

NAME ______ DATE ______

SCORE ______

Performance Task

At the Sports Store

You are working at a sports store. The store sells different kinds of balls.

Write your answers on another piece of paper. Show all your work to receive full credit.

Part A

K.OA.1, K.CC.3

There are 5 soccer balls on the floor. There are 3 more on the shelf. How many are there in all?

5 + 3 = 8

Part B

K.OA.1, K.CC.3

Circle two groups of baseballs to show 8 and 1 more is 9.

Performance Task *(continued)*

Part C

K.OA.1, K.CC.3

Count the number of footballs in each group. Write the numbers. Write how many in all.

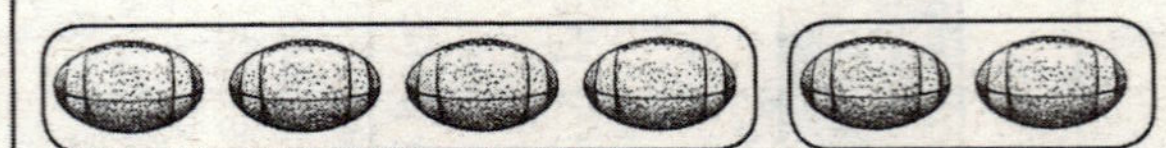

4 + 2 = 6

Part D

K.OA.1, K.CC.3

You sell 12 tennis balls on Monday. You sell 0 tennis balls on Tuesday. How many tennis balls are sold in all? Write the number.

12 + 0 = 12 in all

NAME ______________ DATE ______________

SCORE ______________

Performance Task

At the Sports Store

You are working at a sports store. The store sells different kinds of balls.

Write your answers on another piece of paper. Show all your work to receive full credit.

Part A

K.OA.1, K.CC.3

There are 5 soccer balls on the floor. There are 3 more on the shelf. How many are there in all?

5 + 3 = 7

Part B

K.OA.1, K.CC.3

Circle two groups of baseballs to show 8 and 1 more is 9.

Performance Task *(continued)*

Part C

K.OA.1, K.CC.3

Count the number of footballs in each group. Write the numbers. Write how many in all.

4 + 2 = 6

Part D

K.OA.1, K.CC.3

You sell 12 tennis balls on Monday. You sell 0 tennis balls on Tuesday. How many tennis balls are sold in all? Write the number.

10 + 2 = 12 in all

Student Model

NAME ______________ DATE ______________

SCORE ______________

Performance Task

At the Sports Store

You are working at a sports store. The store sells different kinds of balls.

Write your answers on another piece of paper. Show all your work to receive full credit.

Part A

K.OA.1, K.CC.3

There are 5 soccer balls on the floor. There are 3 more on the shelf. How many are there in all?

5 + 2 = 8

Part B

K.OA.1, K.CC.3

Circle two groups of baseballs to show 8 and 1 more is 9.

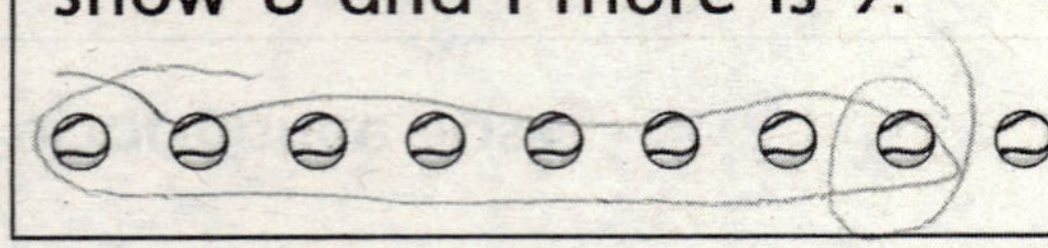

Performance Task *(continued)*

Part C

K.OA.1, K.CC.3

Count the number of footballs in each group. Write the numbers. Write how many in all.

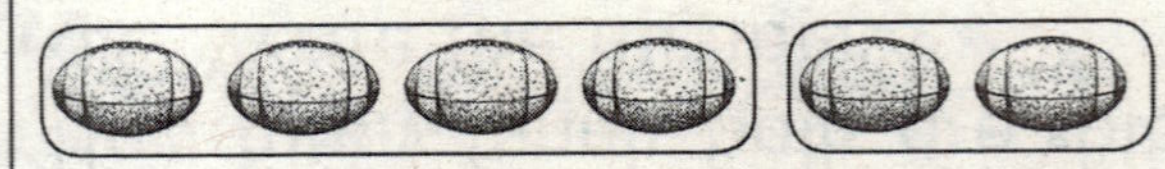

4 + 2 = 5

Part D

K.OA.1, K.CC.3

You sell 12 tennis balls on Monday. You sell 0 tennis balls on Tuesday. How many tennis balls are sold in all? Write the number.

12 + ____ = 12 in all

Copyright © McGraw-Hill Education. Permission is granted to reproduce for classroom use.

NAME ______________________ DATE ____________

SCORE ____________

Performance Task

At the Sports Store

You are working at a sports store. The store sells different kinds of balls.

Write your answers on another piece of paper. Show all your work to receive full credit.

Part A

K.OA.1, K.CC.3

There are 5 soccer balls on the floor. There are 3 more on the shelf. How many are there in all?

5 + 5 = 7

Part B

K.OA.1, K.CC.3

Circle two groups of baseballs to show 8 and 1 more is 9.

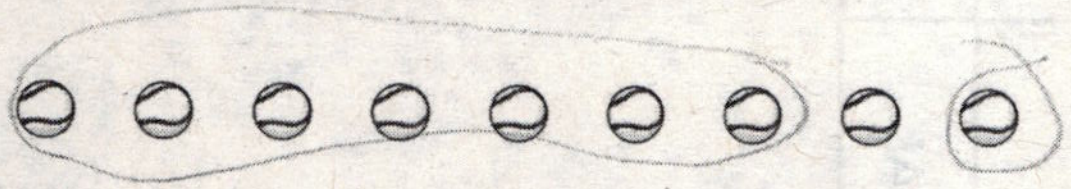

Performance Task *(continued)*

Part C

K.OA.1, K.CC.3

Count the number of footballs in each group. Write the numbers. Write how many in all.

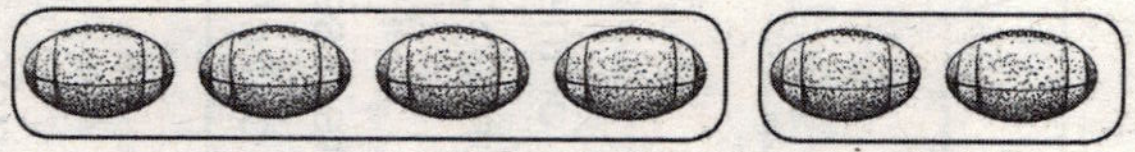

4 + 6 = 2

Part D

K.OA.1, K.CC.3

You sell 12 tennis balls on Monday. You sell 0 tennis balls on Tuesday. How many tennis balls are sold in all? Write the number.

6 + 6 = 12 in all

Student Model

Page 135 • At the Park

Task Scenario		
Students will use numbers through 20 and subtraction to find how many animals are left.		
CCSS Content Standard(s)		K.OA.1, K.CC.3
Mathematical Practices		MP1, MP2, MP4
Depth of Knowledge		DOK2, DOK3
Part	**Maximum Points**	**Scoring Rubric**
A	3	Full Credit: 6 − 3 = 3 Partial Credit (1 point) will be given for the correct subtraction (6 − 3) **OR** a correct answer (3). An additional point will be given for correctly crossing 3 out. No credit will be given for an incorrect sentence AND an incorrect answer AND an incorrect crossing out of 3 objects.
B	1	Full Credit: 3 No credit will be given for an incorrect answer.
C	2	Full Credit: 9 − 5 = 4 Partial Credit (1 point) will be given for the correct subtraction (9 − 5) **OR** a correct answer (4). No credit will be given for an incorrect sentence AND an incorrect answer.
D	1	Full Credit: 10 − 5 = 5 No credit will be given for an incorrect answer.
TOTAL	7	

NAME ______________ DATE ______________

SCORE ______________

Performance Task

At the Park

You count animals while you are at the park.

Write your answers on another piece of paper. Show all your work to receive full credit.

Part A

K.OA.I, K.CC.3

There are six squirrels. 3 run away. Cross out 3 squirrels. How many are left?

6 – 3 = 3

Part B

K.OA.I, K.CC.3

There are 4 cats. One cat leaves. Circle the number that shows how many cats are left.

2 (3) 4

Performance Task *(continued)*

Part C

K.OA.I, K.CC.3

There are 9 birds. Some fly away. Write a number sentence to find how many are left.

9 – 5 = 4

birds left

Part D

K.OA.I, K.CC.3

The squirrels take some acorns. Circle the number sentence that shows the acorns the squirrels took.

(10 – 5 = 5) 9 – 5 = 5 10 – 6 = 4

Student Model

NAME DATE

SCORE

Performance Task

At the Park

You count animals while you are at the park.

Write your answers on another piece of paper. Show all your work to receive full credit.

Part A

K.OA.I, K.CC.3

There are six squirrels. 3 run away. Cross out 3 squirrels. How many are left?

6 – 3 = 2

Part B

K.OA.I, K.CC.3

There are 4 cats. One cat leaves. Circle the number that shows how many cats are left.

2 3 4

Performance Task *(continued)*

Part C

K.OA.I, K.CC.3

There are 9 birds. Some fly away. Write a number sentence to find how many are left.

9 – 5 = 4

birds left

Part D

K.OA.I, K.CC.3

The squirrels take some acorns. Circle the number sentence that shows the acorns the squirrels took.

10 – 5 = 5 9 – 5 = 5 10 – 6 = 4

NAME ______________________ DATE ______________

SCORE ______________

Performance Task

At the Park

You count animals while you are at the park.

Write your answers on another piece of paper. Show all your work to receive full credit.

Part A

K.OA.I, K.CC.3

There are six squirrels. 3 run away. Cross out 3 squirrels. How many are left?

6 – 3 = 3

Part B

K.OA.I, K.CC.3

There are 4 cats. One cat leaves. Circle the number that shows how many cats are left.

2   3   4

Performance Task *(continued)*

Part C

K.OA.I, K.CC.3

There are 9 birds. Some fly away. Write a number sentence to find how many are left.

9 – 4 = 5

birds left

Part D

K.OA.I, K.CC.3

The squirrels take some acorns. Circle the number sentence that shows the acorns the squirrels took.

10 – 5 = 5   9 – 5 = 5   10 – 6 = 4

Student Model

NAME ________ DATE ________

SCORE ________

Performance Task

At the Park

You count animals while you are at the park.

Write your answers on another piece of paper. Show all your work to receive full credit.

Part A

K.OA.1, K.CC.3

There are six squirrels. 3 run away. Cross out 3 squirrels. How many are left?

6 – 2 = 4

Part B

K.OA.1, K.CC.3

There are 4 cats. One cat leaves. Circle the number that shows how many cats are left.

2   3   4

Performance Task *(continued)*

Part C

K.OA.1, K.CC.3

There are 9 birds. Some fly away. Write a number sentence to find how many are left.

9 – 4 = 3

birds left

Part D

K.OA.1, K.CC.3

The squirrels take some acorns. Circle the number sentence that shows the acorns the squirrels took.

$10 - 5 = 5$ $9 - 5 = 5$ $10 - 6 = 4$

Page 137 • At the Airport

Task Scenario		
Students will use numbers 11 through 19 to add and subtract and count airplanes.		
CCSS Content Standard(s)		K.OA.1, K.NBT.1
Mathematical Practices		MP1, MP2, MP4
Depth of Knowledge		DOK2, DOK3, DOK4
Part	**Maximum Points**	**Scoring Rubric**
A	2	Full Credit: 10 + 4 = 14 Partial Credit (1 point) will be given for the correct parts (10 and 4) **OR** the correct total (14). No credit will be given for incorrect parts AND an incorrect total.
B	2	Full Credit: 10; 2 airplanes crossed out Partial Credit (1 point) will be given for having 2 airplanes crossed out **OR** having a correct answer (10). No credit will be given for incorrectly crossing out AND an incorrect answer.
C	2	Full Credit: 17 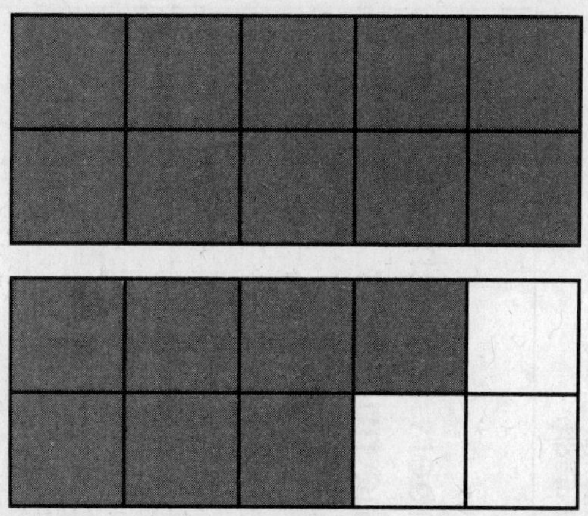 Partial Credit will be given for a correct shading **OR** a correct answer (17). No credit will be given for an incorrect answer AND incorrect shading.
TOTAL	6	

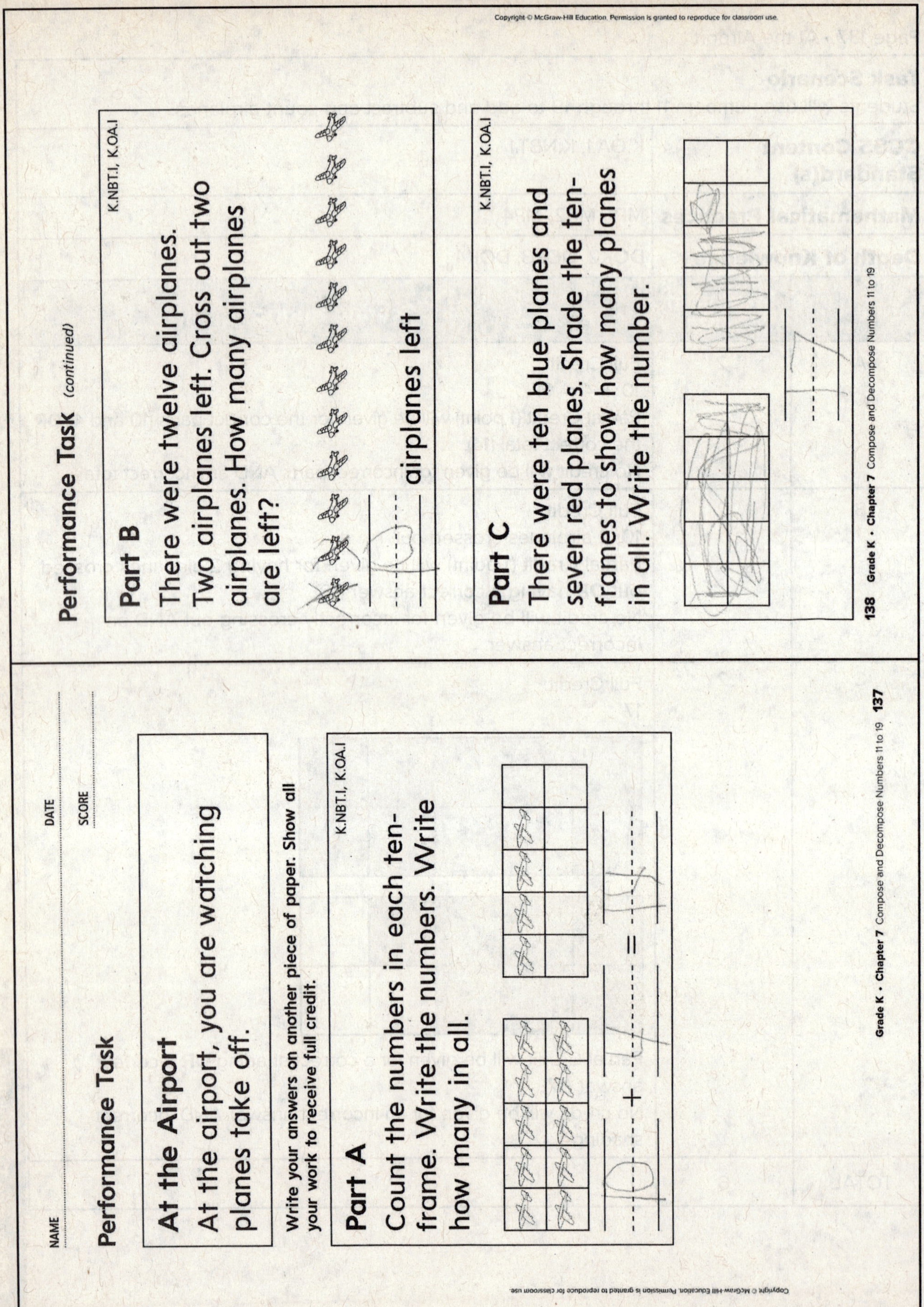
Performance Task
NAME
DATE
SCORE
At the Airport
At the airport, you are watching planes take off.
Write your answers on another piece of paper. Show all your work to receive full credit.
Part A
K.NBT.1, K.OA.1
Count the numbers in each ten-frame. Write the numbers. Write how many in all.
10 + 4 = 14
Copyright © McGraw-Hill Education. Permission is granted to reproduce for classroom use.
Grade K • Chapter 7 Compose and Decompose Numbers 11 to 19 137
Performance Task (continued)
Part B
K.NBT.1, K.OA.1
There were twelve airplanes. Two airplanes left. Cross out two airplanes. How many airplanes are left?
airplanes left
Part C
K.NBT.1, K.OA.1
There were ten blue planes and seven red planes. Shade the ten-frames to show how many planes in all. Write the number.
17
138 Grade K • Chapter 7 Compose and Decompose Numbers 11 to 19
Copyright © McGraw-Hill Education. Permission is granted to reproduce for classroom use.

NAME ______________ DATE ______________

SCORE ______________

Performance Task

At the Airport

At the airport, you are watching planes take off.

Write your answers on another piece of paper. Show all your work to receive full credit.

Part A

K.NBT.1, K.OA.1

Count the numbers in each ten-frame. Write the numbers. Write how many in all.

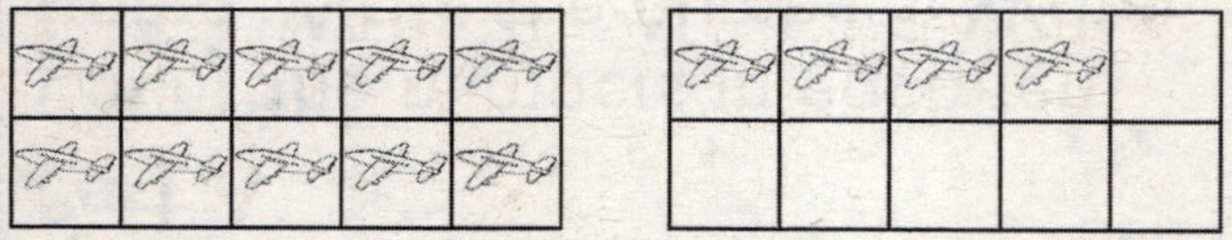

10 + 4 = 4

Performance Task *(continued)*

Part B

K.NBT.1, K.OA.1

There were twelve airplanes. Two airplanes left. Cross out two airplanes. How many airplanes are left?

10 airplanes left

Part C

K.NBT.1, K.OA.1

There were ten blue planes and seven red planes. Shade the ten-frames to show how many planes in all. Write the number.

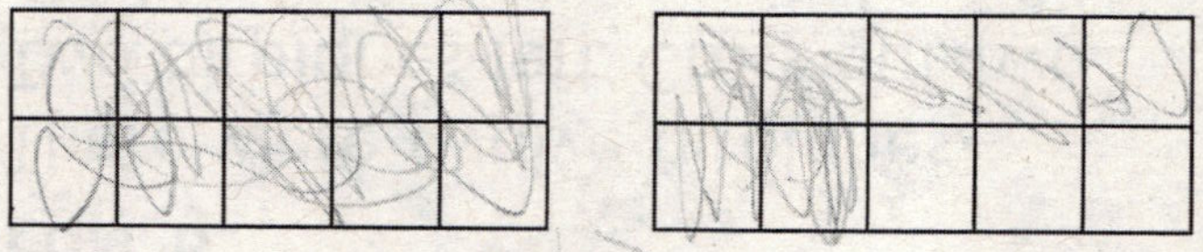

7

Student Model

NAME ____________ DATE ____________

SCORE ____________

Performance Task

At the Airport

At the airport, you are watching planes take off.

Write your answers on another piece of paper. Show all your work to receive full credit.

Part A

K.NBT.1, K.OA.1

Count the numbers in each ten-frame. Write the numbers. Write how many in all.

____ + ____ = ____

Performance Task *(continued)*

Part B

K.NBT.1, K.OA.1

There were twelve airplanes. Two airplanes left. Cross out two airplanes. How many airplanes are left?

____ airplanes left

Part C

K.NBT.1, K.OA.1

There were ten blue planes and seven red planes. Shade the ten-frames to show how many planes in all. Write the number.

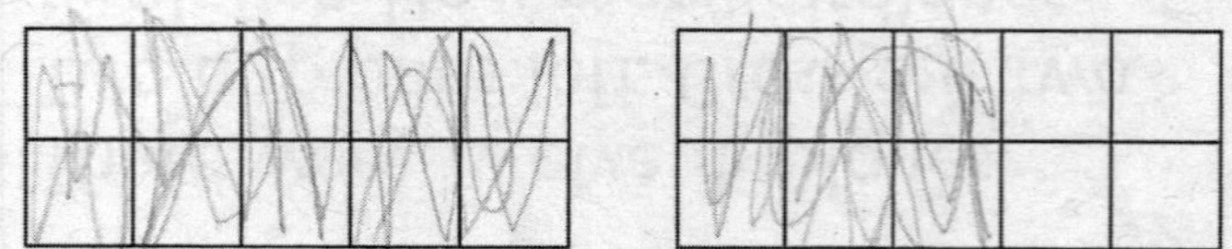

____

NAME ____________ DATE ____________

SCORE ____________

Performance Task

At the Airport

At the airport, you are watching planes take off.

Write your answers on another piece of paper. Show all your work to receive full credit.

Part A

K.NBT.1, K.OA.1

Count the numbers in each ten-frame. Write the numbers. Write how many in all.

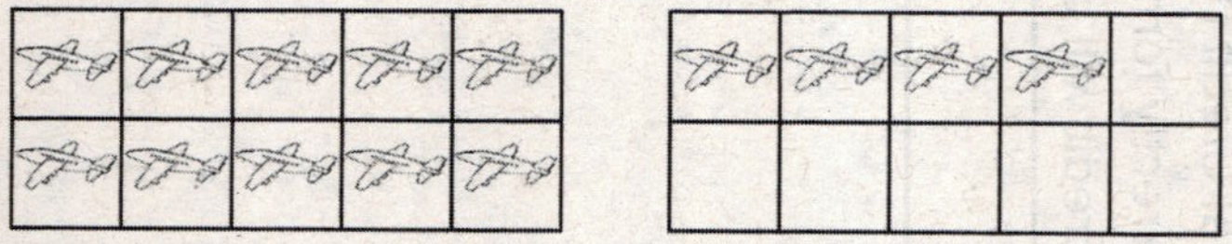

10 + 3 = 13

Performance Task *(continued)*

Part B

K.NBT.1, K.OA.1

There were twelve airplanes. Two airplanes left. Cross out two airplanes. How many airplanes are left?

10 airplanes left

Part C

K.NBT.1, K.OA.1

There were ten blue planes and seven red planes. Shade the ten-frames to show how many planes in all. Write the number.

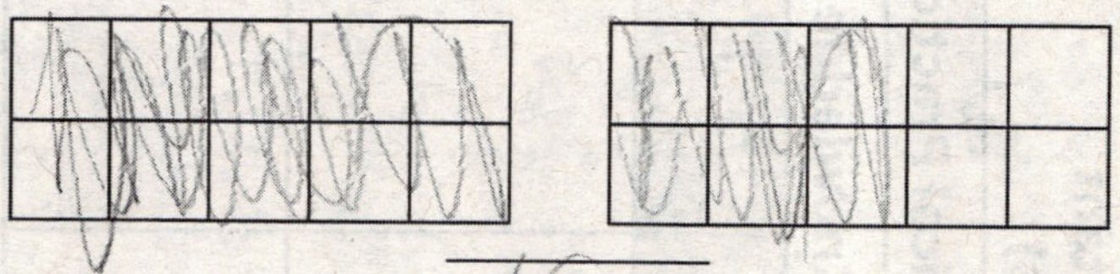

16

Student Model

Page 139 • Baking Bread

Task Scenario	
Students will use measurements of length, capacity, and weight to compare kitchen objects.	
CCSS Content Standard(s)	K.MD.2, K.CC.3, K.CC.4
Mathematical Practices	MP1, MP2, MP4, MP5
Depth of Knowledge	DOK2, DOK3

Part	Maximum Points	Scoring Rubric
A	2	Full Credit: Student circles the larger bowl and crosses out the smaller bowl. Partial Credit (1 point) will be given for the correct circle **OR** the correct cross out. No credit will be given for 2 incorrect responses.
B	1	Full Credit: Student correctly circles the one loaf of bread that is longer than the tray. No credit will be given for an incorrect choice.
C	1	Full Credit: Student correctly crosses out the table and nothing else. No credit will be given for an incorrect choice.
D	2	Full Credit: 17 Partial Credit (1 point) will be given for a correct answer (17) but an incorrectly formed number. No credit will be given for an incorrect answer.
TOTAL	6	

NAME ______________ DATE ______________

SCORE ______________

Performance Task

Baking Bread

You are baking bread with your mother.

Write your answers on another piece of paper. Show all your work to receive full credit.

Part A

K.MD.2

Circle the mixing bowl that holds more. Cross out the one that holds less.

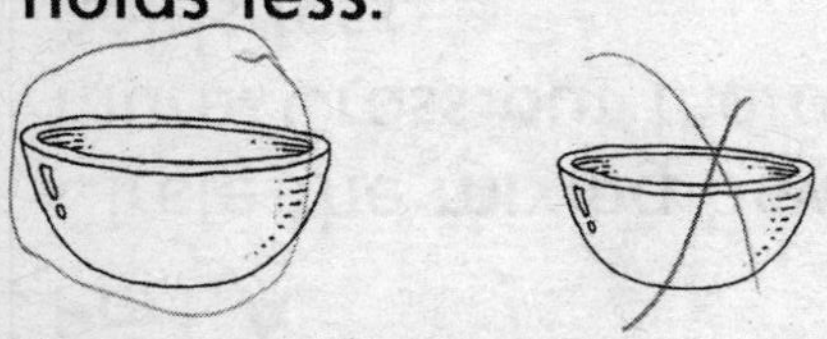

Part B

K.MD.2

Circle the loaf of bread that is longer than the tray.

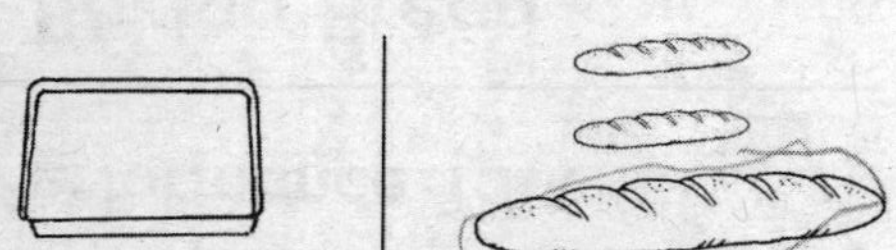

Performance Task *(continued)*

Part C

K.MD.2

Cross out the kitchen object that is too heavy to lift by yourself.

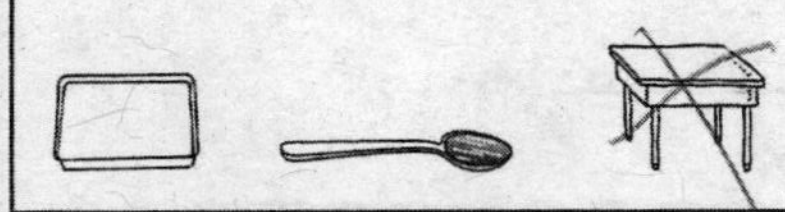

Part D

K.MD.2, K.CC.3, K.CC.4

How many pennies long is the loaf of bread? Write the number.

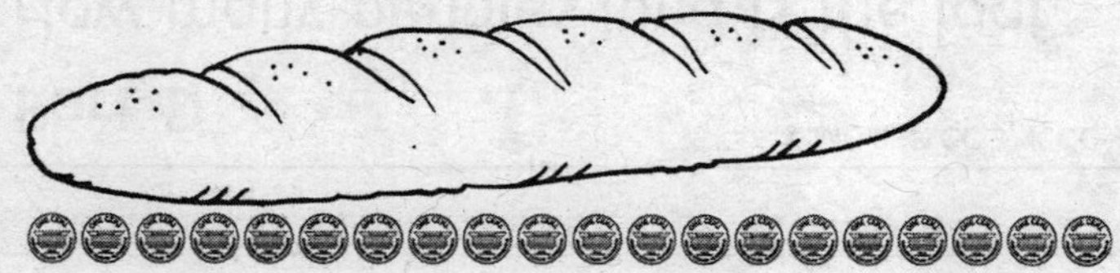

17

Student Model

NAME DATE

SCORE

Performance Task

Baking Bread

You are baking bread with your mother.

Write your answers on another piece of paper. Show all your work to receive full credit.

Part A

K.MD.2

Circle the mixing bowl that holds more. Cross out the one that holds less.

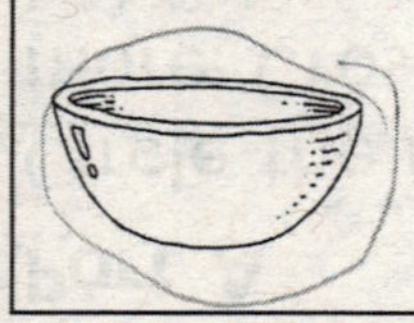

Part B

K.MD.2

Circle the loaf of bread that is longer than the tray.

Performance Task *(continued)*

Part C

K.MD.2

Cross out the kitchen object that is too heavy to lift by yourself.

Part D

K.MD.2, K.CC.3, K.CC.4

How many pennies long is the loaf of bread? Write the number.

17

NAME ______________________ DATE ____________

SCORE ____________

Performance Task

Baking Bread

You are baking bread with your mother.

Write your answers on another piece of paper. Show all your work to receive full credit.

Part A

K.MD.2

Circle the mixing bowl that holds more. Cross out the one that holds less.

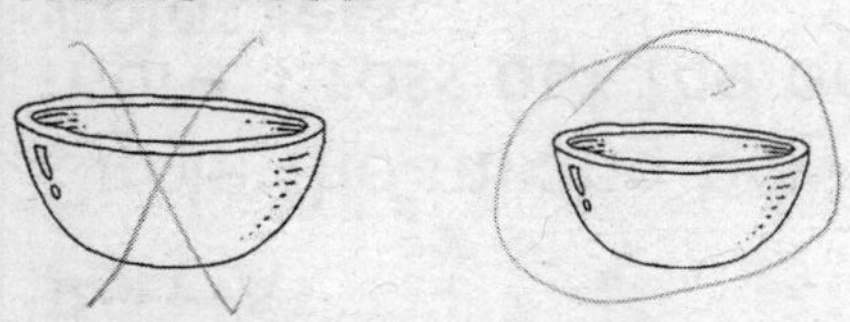

Part B

K.MD.2

Circle the loaf of bread that is longer than the tray.

Performance Task *(continued)*

Part C

K.MD.2

Cross out the kitchen object that is too heavy to lift by yourself.

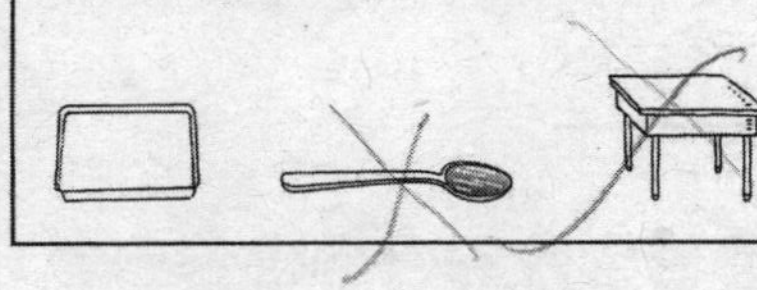

Part D

K.MD.2, K.CC.3, K.CC.4

How many pennies long is the loaf of bread? Write the number.

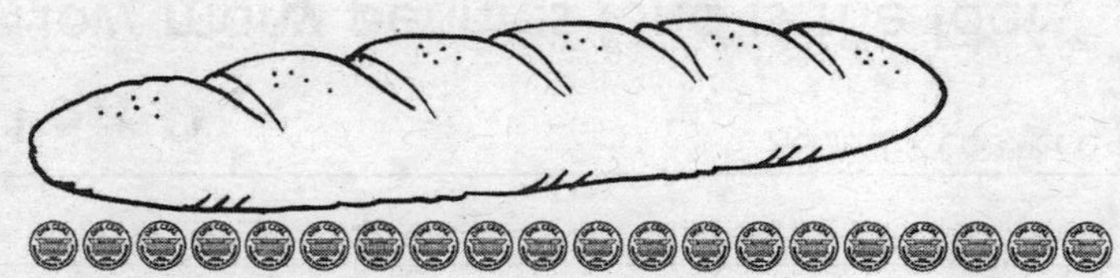

Student Model

NAME ______________ DATE ______________

SCORE ______________

Performance Task

Baking Bread

You are baking bread with your mother.

Write your answers on another piece of paper. Show all your work to receive full credit.

Part A

K.MD.2

Circle the mixing bowl that holds more. Cross out the one that holds less.

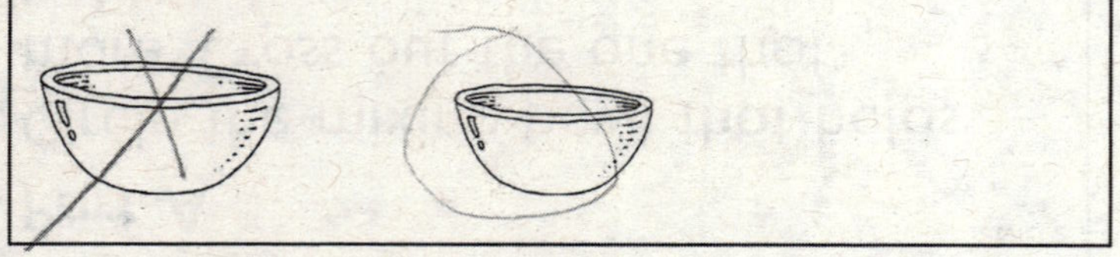

Part B

K.MD.2

Circle the loaf of bread that is longer than the tray.

Performance Task *(continued)*

Part C

K.MD.2

Cross out the kitchen object that is too heavy to lift by yourself.

Part D

K.MD.2, K.CC.3, K.CC.4

How many pennies long is the loaf of bread? Write the number.

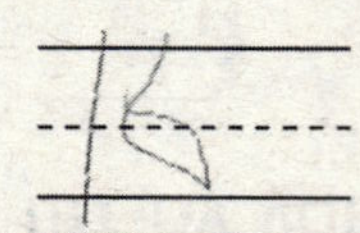

Page 141 • A Trip to the Grocery Store

Task Scenario Students will compare and contrast objects to classify and describe them in terms of their number, size, and shape.	
CCSS Content Standard(s)	K.MD.3, K.CC.3
Mathematical Practices	MP1, MP2, MP4
Depth of Knowledge	DOK2, DOK3, DOK4

Part	Maximum Points	Scoring Rubric
A	1	Full Credit: Student circles the three carrots that are alike. No credit will be given for an incorrect answer.
B	2	Full Credit: Student crosses out the chalk and the chisel. Partial Credit (1 point) will be given for crossing out the chalk **OR** the chisel. No credit will be given for crossing out an edible object.
C	1	Full Credit: Student underlines each of the three larger muffins No credit will be given for an incorrect answer.
D	1	Full Credit: Student circles the three cans. No credit will be given for an incorrect answer.
E	3	Full Credit: Student circles the eggs and the cherries, but not the flowers. Partial Credit (1 points) will be awarded for each object being circled or not circled as they should be. No credit will be given for three incorrect answers (eggs and cherries are not circled, and flowers are).
TOTAL	8	

NAME ____________________ DATE ____________

SCORE ____________

Performance Task

A Trip to the Grocery Store

You and your father take a trip to the grocery store. Research how many objects are in a dozen.

Write your answers on another piece of paper. Show all your work to receive full credit.

Part A

K.MD.3

Circle the carrots that are alike.

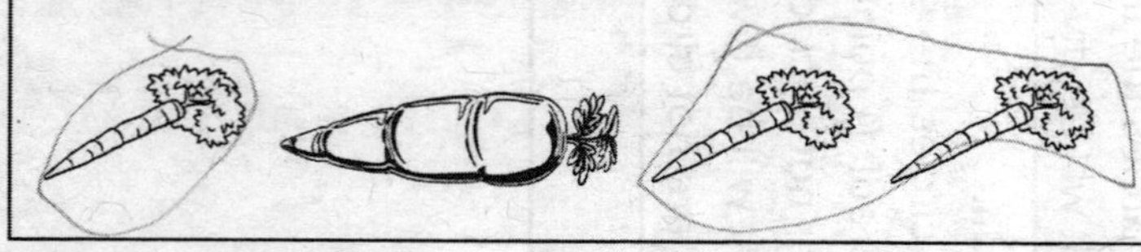

Part B

K.MD.3

Cross out the items that you cannot eat.

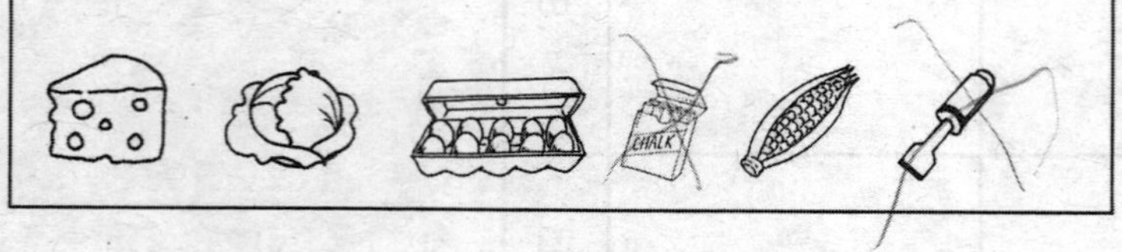

Performance Task *(continued)*

Part C

K.MD.3

Sort by size. Underline the larger muffins.

Part D

K.MD.3

Sort by shape. Circle all the ones that are alike.

Part E

K.MD.3, K.CC.3

Sort by number. Circle the objects that have one dozen.

NAME ________________ DATE ________

SCORE ________

Performance Task

A Trip to the Grocery Store

You and your father take a trip to the grocery store. Research how many objects are in a dozen.

Write your answers on another piece of paper. Show all your work to receive full credit.

Part A
K.MD.3

Circle the carrots that are alike.

Part B
K.MD.3

Cross out the items that you cannot eat.

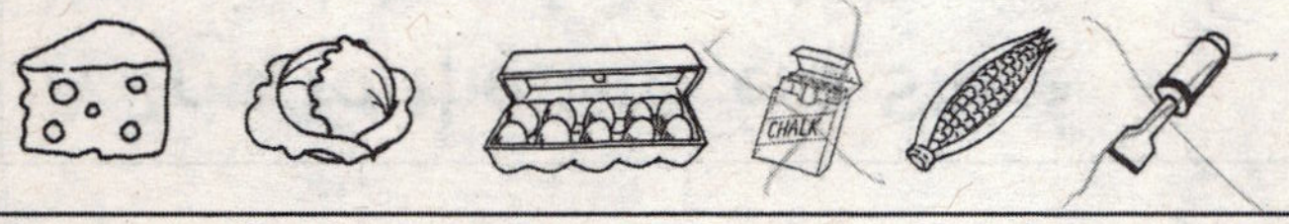

Performance Task *(continued)*

Part C
K.MD.3

Sort by size. Underline the larger muffins.

Part D
K.MD.3

Sort by shape. Circle all the ones that are alike.

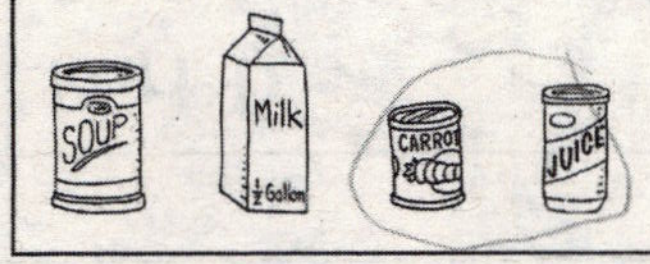

Part E
K.MD.3, K.CC.3

Sort by number. Circle the objects that have one dozen.

Student Model

NAME ______________ DATE ______________

SCORE ______________

Performance Task

A Trip to the Grocery Store

You and your father take a trip to the grocery store. Research how many objects are in a dozen.

Write your answers on another piece of paper. Show all your work to receive full credit.

Part A

K.MD.3

Circle the carrots that are alike.

Part B

K.MD.3

Cross out the items that you cannot eat.

Performance Task *(continued)*

Part C

K.MD.3

Sort by size. Underline the larger muffins.

Part D

K.MD.3

Sort by shape. Circle all the ones that are alike.

Part E

K.MD.3, K.CC.3

Sort by number. Circle the objects that have one dozen.

NAME ____________ DATE ____________

SCORE ____________

Performance Task

A Trip to the Grocery Store

You and your father take a trip to the grocery store. Research how many objects are in a dozen.

Write your answers on another piece of paper. Show all your work to receive full credit.

Part A

K.MD.3

Circle the carrots that are alike.

Part B

K.MD.3

Cross out the items that you cannot eat.

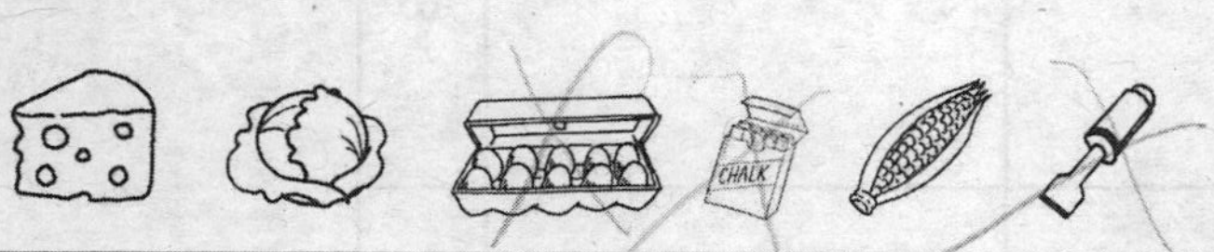

Performance Task *(continued)*

Part C

K.MD.3

Sort by size. Underline the larger muffins.

Part D

K.MD.3

Sort by shape. Circle all the ones that are alike.

Part E

K.MD.3, K.CC.3

Sort by number. Circle the objects that have one dozen.

Student Model

Page 143 • A Scavenger Hunt

Task Scenario	
Students will use relational terms like "above", "below", "behind", "in front of", and "near", as well as counting and subtraction to answer questions about objects from a scavenger hunt.	
CCSS Content Standard(s)	K.G.1, K.OA.1, K.CC.3
Mathematical Practices	MP1, MP2, MP4, MP5
Depth of Knowledge	DOK2, DOK3, DOK4

Part	Maximum Points	Scoring Rubric
A	2	Full Credit: 3 Student correctly circles all the objects under the table. Partial Credit (1 point) will be given for a correct answer (3) **OR** a correct circle. No credit will be given for an incorrect answer AND an incorrect circle.
B	2	Full Credit: Student underlines the feather nearest to the duck. No credit will be given for an incorrect answer.
C	2	Full Credit: 2 Student correctly circles all objects that are behind the teepee. Partial Credit (1 point) will be given for a correct answer (2) **OR** a correct circle. No credit will be given for an incorrect answer AND an incorrect circle.
D	2	Full Credit: Student correctly crosses out 3 apples. 2 Partial Credit (1 point) will be given for a correct answer (2) **OR** a correct crossing out. No credit will be given for an incorrect answer AND an incorrect crossing out.
TOTAL	8	

NAME ________ DATE ________

SCORE ________

Performance Task

A Scavenger Hunt

You are looking for some objects for a scavenger hunt.

Write your answers on another piece of paper. Show all your work to receive full credit.

Part A

K.G.I, K.CC.3

How many objects are below the table? Circle them. Write the number.

3

Part B

K.G.I, K.CC.3

Underline the feather that is nearest to the duck.

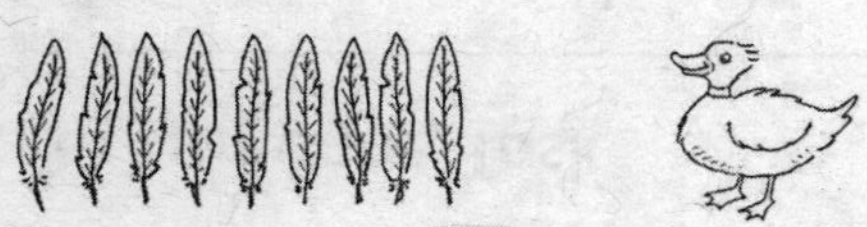

Performance Task *(continued)*

Part C

K.G.I, K.CC.3

How many objects are behind the teepee? Circle them. Write the number.

2

Part D

K.G.I, K.CC.3, K.OA.I

There are some objects above the desk. You take away three of the objects. Cross out three of the objects above the desk. How many are left above the desk? Write the number.

2

Student Model

NAME ________________ DATE ________

SCORE ________

Performance Task

A Scavenger Hunt

You are looking for some objects for a scavenger hunt.

Write your answers on another piece of paper. Show all your work to receive full credit.

Part A

K.G.1, K.CC.3

How many objects are below the table? Circle them. Write the number.

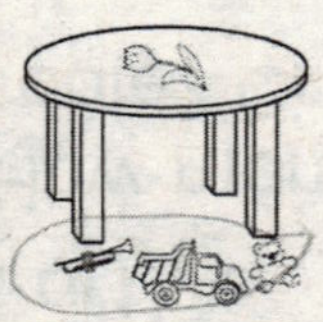

2

Part B

K.G.1, K.CC.3

Underline the feather that is nearest to the duck.

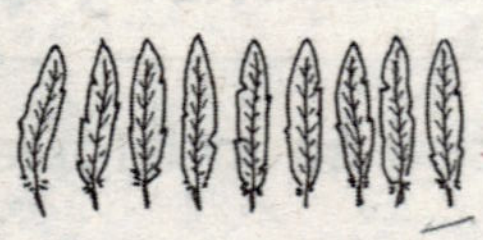

Performance Task *(continued)*

Part C

K.G.1, K.CC.3

How many objects are behind the teepee? Circle them. Write the number.

1

Part D

K.G.1, K.CC.3, K.OA.1

There are some objects above the desk. You take away three of the objects. Cross out three of the objects above the desk. How many are left above the desk? Write the number.

1

NAME ______________________ DATE ______________

SCORE ______________

Performance Task

A Scavenger Hunt

You are looking for some objects for a scavenger hunt.

Write your answers on another piece of paper. Show all your work to receive full credit.

Part A

K.G.1, K.CC.3

How many objects are below the table? Circle them. Write the number.

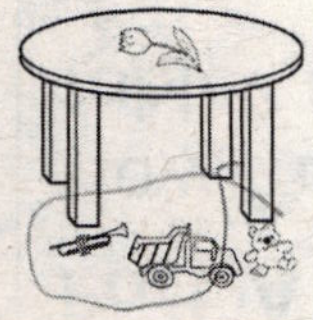

Part B

K.G.1, K.CC.3

Underline the feather that is nearest to the duck.

Performance Task *(continued)*

Part C

K.G.1, K.CC.3

How many objects are behind the teepee? Circle them. Write the number.

Part D

K.G.1, K.CC.3, K.OA.1

There are some objects above the desk. You take away three of the objects. Cross out three of the objects above the desk. How many are left above the desk? Write the number.

Student Model

NAME ______________ DATE ______________

SCORE ______________

Performance Task

A Scavenger Hunt

You are looking for some objects for a scavenger hunt.

Write your answers on another piece of paper. Show all your work to receive full credit.

Part A

K.G.I, K.CC.3

How many objects are below the table? Circle them. Write the number.

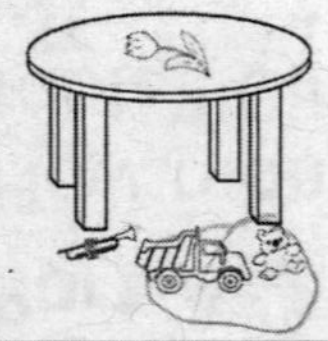

Part B

K.G.I, K.CC.3

Underline the feather that is nearest to the duck.

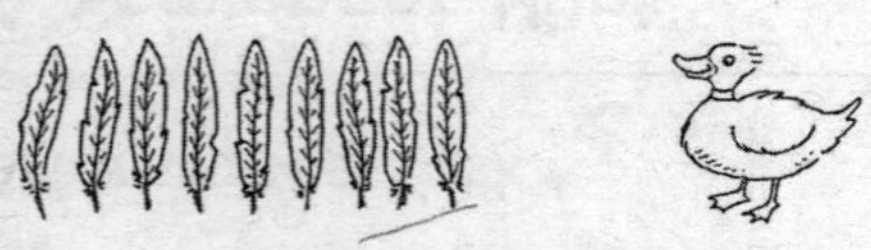

Performance Task *(continued)*

Part C

K.G.I, K.CC.3

How many objects are behind the teepee? Circle them. Write the number.

Part D

K.G.I, K.CC.3, K.OA.I

There are some objects above the desk. You take away three of the objects. Cross out three of the objects above the desk. How many are left above the desk? Write the number.

Page 145 • Building with Tiles

Task Scenario Students will use squares, hexagons, and triangles, identify and count the shapes, and complete patterns and make composite shapes using two-dimensional figures.			
CCSS Content Standard(s)		K.G.2, K.G.4, K.G.5, K.CC.3, K.OA.4	
Mathematical Practices		MP1, MP2, MP4, MP7, MP8	
Depth of Knowledge		DOK2, DOK3, DOK4	
Part	**Maximum Points**	**Scoring Rubric**	
A	3	Full Credit: Student circles the triangle, underlines the hexagon, and crosses out the square. Partial Credit (1 point) will be given for each correct marking. No credit will be given for three incorrect markings.	
B	2	Full Credit: Student correct draws a square in the first blank and a triangle in the second blank. Partial Credit (1 point) will be given for each correct response. No credit will be given for two incorrect shapes.	
C	3	Full Credit: 0, 6, 1 Partial Credit (1 point) will be given for each correct response. No credit will be given for three incorrect responses.	
D	2	Full Credit: 2 Student correctly crosses out 8 triangles. Partial Credit (1 point) will be given for a correct answer (2) **OR** for correctly crossing out 8 triangles. No credit will be given for an incorrect answer AND an incorrect crossing out of 8 triangles.	
TOTAL	10		

NAME

DATE

SCORE

Performance Task

Building with Tiles

Marcello has tiles that are different shapes.

Write your answers on another piece of paper. Show all your work to receive full credit.

Part A

K.G.2

The different tiles are shown. Circle the triangle. Underline the hexagon. Put an "X" on the square.

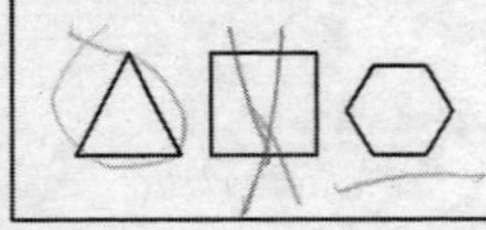

Part B

K.G.4

Marcello makes a pattern with his tiles. Draw the two tiles that come next.

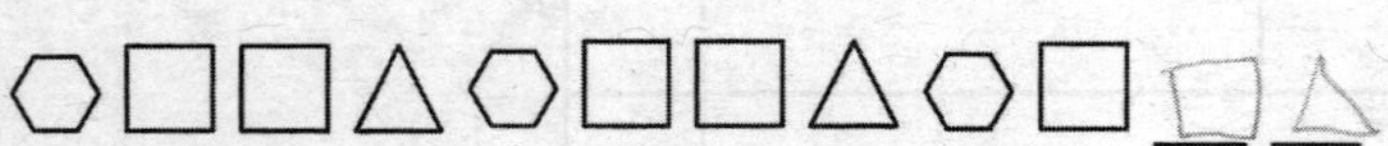

Performance Task *(continued)*

Part C

K.G.5, K.CC.3

Marcello builds a star design out of tiles. How many of each tile does he use? Write the numbers.

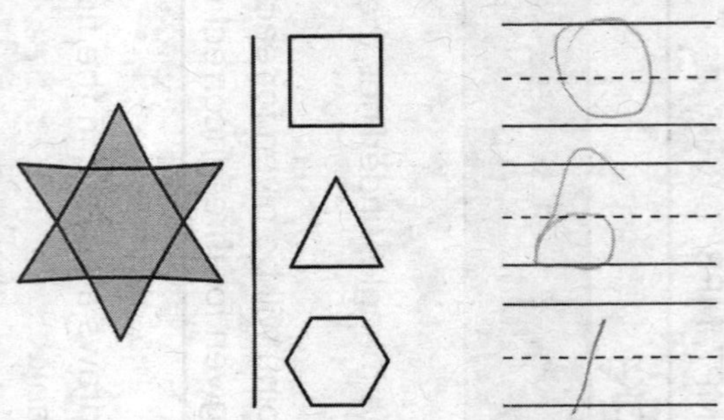

Part D

K.G.2, K.OA.4

Marcello has these shapes. He uses 8 of his triangles in another design. Cross out 8 triangles. How many triangles are left? Write the number.

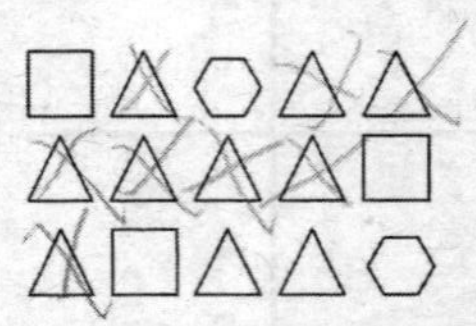

2 triangles left

Copyright © McGraw-Hill Education. Permission is granted to reproduce for classroom use.

NAME DATE

SCORE

Performance Task

Building with Tiles

Marcello has tiles that are different shapes.

Write your answers on another piece of paper. Show all your work to receive full credit.

Part A

K.G.2

The different tiles are shown. Circle the triangle. Underline the hexagon. Put an "X" on the square.

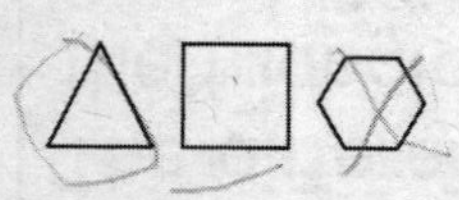

Part B

K.G.4

Marcello makes a pattern with his tiles. Draw the two tiles that come next.

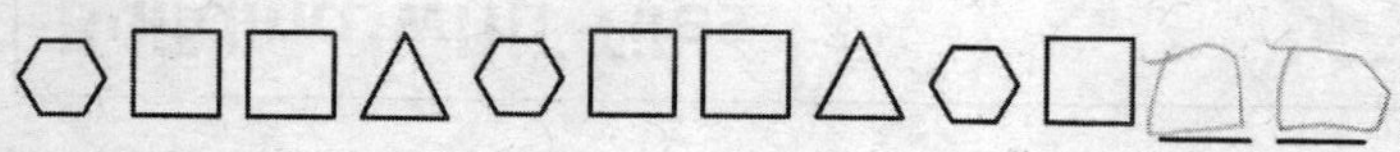

Performance Task *(continued)*

Part C

K.G.5, K.CC.3

Marcello builds a star design out of tiles. How many of each tile does he use? Write the numbers.

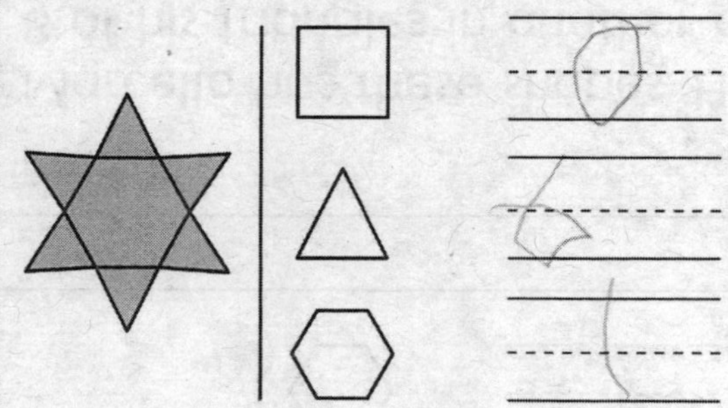

Part D

K.G.2, K.OA.4

Marcello has these shapes. He uses 8 of his triangles in another design. Cross out 8 triangles. How many triangles are left? Write the number.

2 triangles left

Student Model

NAME ________________ DATE ________________

SCORE ________________

Performance Task

Building with Tiles

Marcello has tiles that are different shapes.

Write your answers on another piece of paper. Show all your work to receive full credit.

Part A

K.G.2

The different tiles are shown. Circle the triangle. Underline the hexagon. Put an "X" on the square.

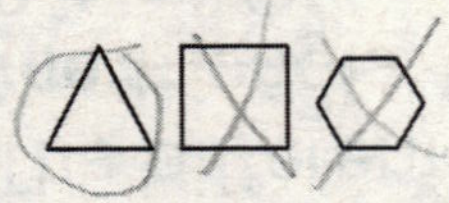

Part B

K.G.4

Marcello makes a pattern with his tiles. Draw the two tiles that come next.

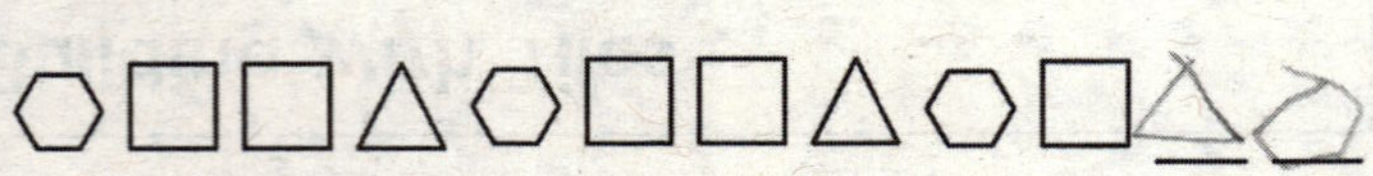

Performance Task *(continued)*

Part C

K.G.5, K.CC.3

Marcello builds a star design out of tiles. How many of each tile does he use? Write the numbers.

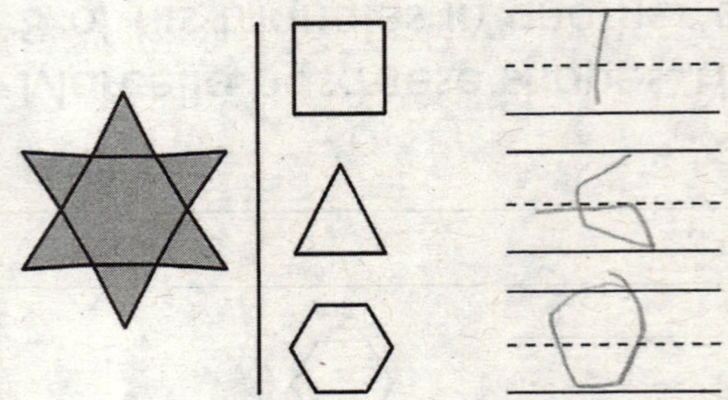

Part D

K.G.2, K.OA.4

Marcello has these shapes. He uses 8 of his triangles in another design. Cross out 8 triangles. How many triangles are left? Write the number.

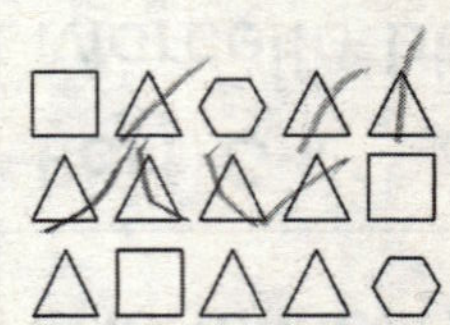

________ triangles left

NAME ________________ DATE ________________

SCORE ________________

Performance Task

Building with Tiles

Marcello has tiles that are different shapes.

Write your answers on another piece of paper. Show all your work to receive full credit.

Part A

K.G.2

The different tiles are shown. Circle the triangle. Underline the hexagon. Put an "X" on the square.

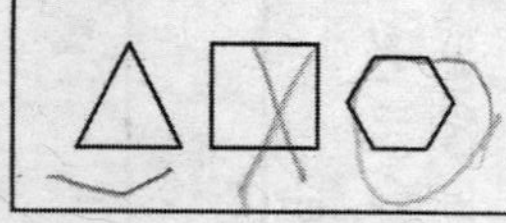

Part B

K.G.4

Marcello makes a pattern with his tiles. Draw the two tiles that come next.

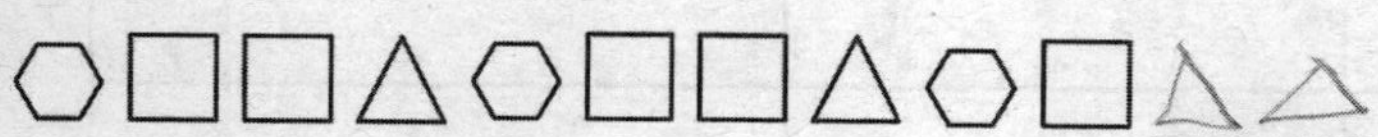

Performance Task *(continued)*

Part C

K.G.5, K.CC.3

Marcello builds a star design out of tiles. How many of each tile does he use? Write the numbers.

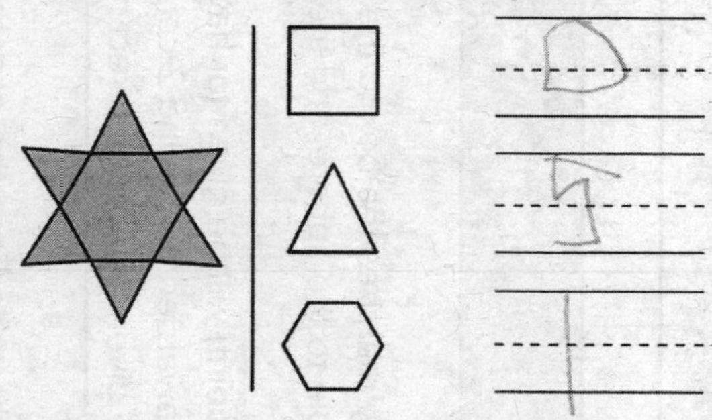

Part D

K.G.2, K.OA.4

Marcello has these shapes. He uses 8 of his triangles in another design. Cross out 8 triangles. How many triangles are left? Write the number.

______ triangles left

Student Model

Page 147 • Cleaning Out the Pantry

Task Scenario Students will use cones, cubes, cylinders, and spheres to describe and classify real world objects.		
CCSS Content Standard(s)		K.G.2, K.G.4, K.CC.3
Mathematical Practices		MP1, MP2, MP4
Depth of Knowledge		DOK2, DOK3
Part	**Maximum Points**	**Scoring Rubric**
A	3	Full Credit: Student correctly matches the box to the cube, the can to the cylinder, the apple to the sphere, and the ice cream cone to the cone. Partial Credit (1 point) will be given for having 1 correct match. 2 points will be awarded for having 2 correct matches. No credit will be given for four incorrect matches.
B	2	Full Credit: Student correctly underlines the can, the cone, and the apple. 3 Partial Credit (1 point) will be given for having the correct answer (3) **OR** the correct underlines. No credit will be given for an incorrect answer AND incorrect underlines.
C	2	Full Credit: Student correctly circles the can and the crackers. 2 Partial Credit (1 point) will be given for having the correct answer (2) **OR** the correct circles. No credit will be given for an incorrect answer AND incorrect circles.
D	2	Full Credit: Student correctly underlines the can, the cone, and the crackers. 3 Partial Credit (1 point) will be given for having the correct answer (3) **OR** the correct underlines. No credit will be given for an incorrect answer AND incorrect underlines.
TOTAL	9	

NAME ________________ DATE ________________

SCORE ________________

Performance Task

Cleaning Out the Pantry

There are several objects in Mrs. Hernandez's pantry.

Write your answers on another piece of paper. Show all your work to receive full credit.

Part A

K.G.2

Match each object to the three-dimensional shape.

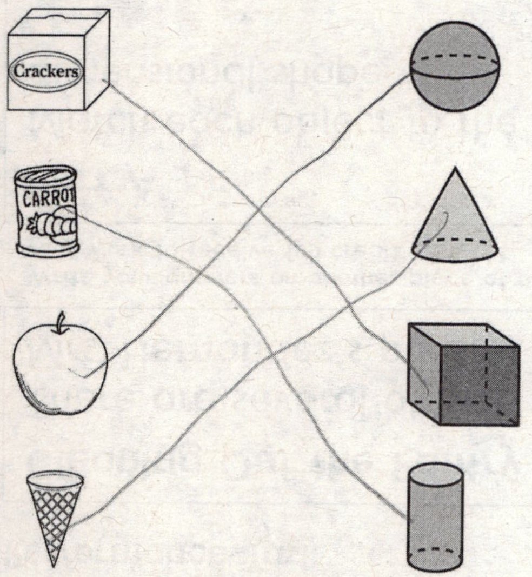

Performance Task *(continued)*

Part B

K.G.4, K.CC.3

Underline the shapes that will roll. Count them. Write the number.

Part C

K.G.4, K.CC.3

Circle the shapes that will stack. Count them. Write the number.

Part D

K.G.4, K.CC.3

Underline the shapes that will slide. Count them. Write the number.

Student Model

NAME ______________________ DATE ____________

SCORE ____________

Performance Task

Cleaning Out the Pantry

There are several objects in Mrs. Hernandez's pantry.

Write your answers on another piece of paper. Show all your work to receive full credit.

Part A

K.G.2

Match each object to the three-dimensional shape.

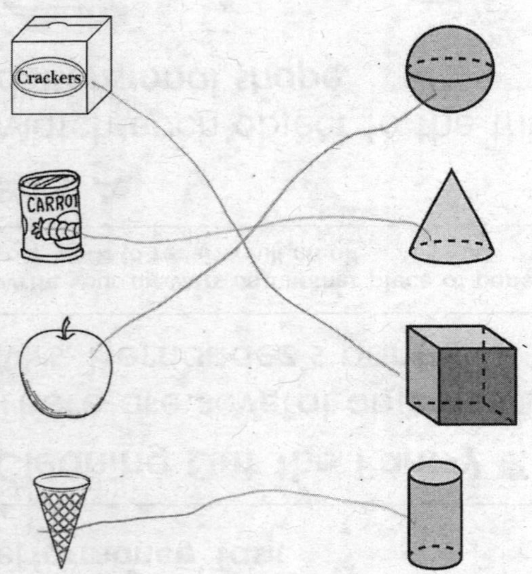

Performance Task *(continued)*

Part B

K.G.4, K.CC.3

Underline the shapes that will roll. Count them. Write the number.

Part C

K.G.4, K.CC.3

Circle the shapes that will stack. Count them. Write the number.

Part D

K.G.4, K.CC.3

Underline the shapes that will slide. Count them. Write the number.

NAME ______________ DATE ______________

SCORE ______________

Performance Task

Cleaning Out the Pantry

There are several objects in Mrs. Hernandez's pantry.

Write your answers on another piece of paper. Show all your work to receive full credit.

Part A

K.G.2

Match each object to the three-dimensional shape.

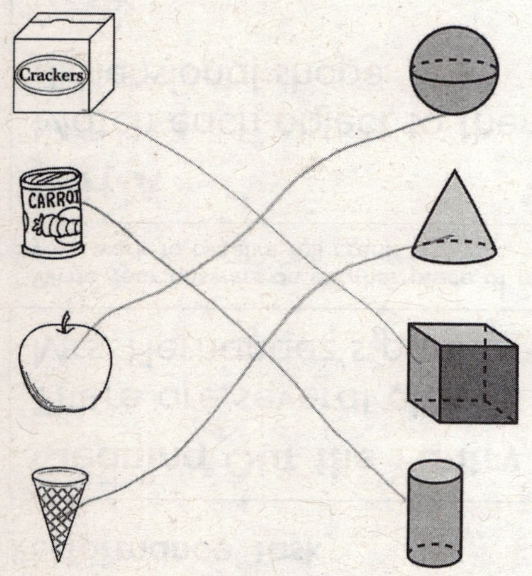

Performance Task *(continued)*

Part B

K.G.4, K.CC.3

Underline the shapes that will roll. Count them. Write the number.

Part C

K.G.4, K.CC.3

Circle the shapes that will stack. Count them. Write the number.

Part D

K.G.4, K.CC.3

Underline the shapes that will slide. Count them. Write the number.

Student Model

NAME ____________ DATE ________

SCORE ________

Performance Task

Cleaning Out the Pantry

There are several objects in Mrs. Hernandez's pantry.

Write your answers on another piece of paper. Show all your work to receive full credit.

Part A

K.G.2

Match each object to the three-dimensional shape.

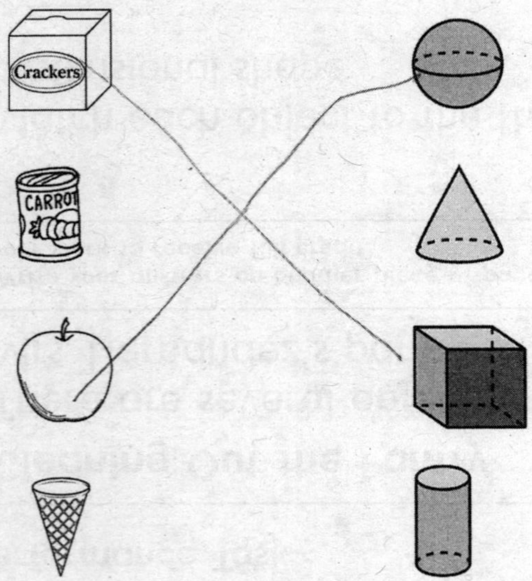

Performance Task *(continued)*

Part B

K.G.4, K.CC.3

Underline the shapes that will roll. Count them. Write the number.

Part C

K.G.4, K.CC.3

Circle the shapes that will stack. Count them. Write the number.

Part D

K.G.4, K.CC.3

Underline the shapes that will slide. Count them. Write the number.

NAME ______________ DATE ______________

Benchmark Test I

SCORE ______________

1. Circle 18 coats. K.CC.3

2. Count the horseshoes. Write the number above the horseshoes. K.OA.3

9

Show a way to take apart the number.

Sample answer:

5 and 4

3. Circle the number that is one more than 7. K.CC.4c

6 7 8 9

4. Circle the basket with zero apples. K.CC.6

5. Connect the numbers in order. Circle the largest number. K.CC.1

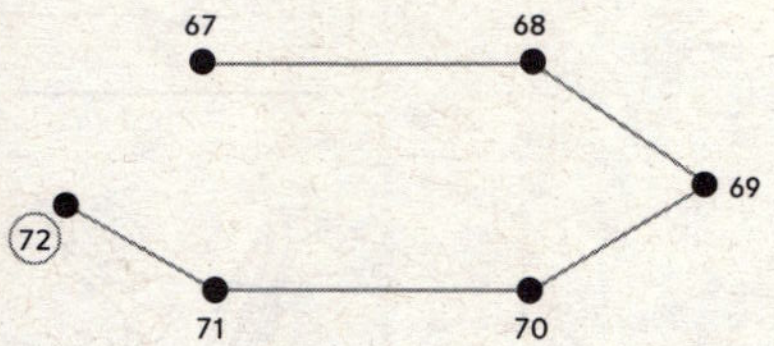

6. Show a way to make 7 by circling houses. Write the numbers. K.OA.1

Sample answer:

2 and 5

Benchmark Tests

7. Count the dots. Draw more dots to show nine. Write the number. K.CC.3, K.CC.4

○ ○ ○ ○ ○ ○ ○ ○ ○

9

8. Circle the deer that is first in line to go to the pond. Draw an X on the deer that is third in line. Draw a line under the deer that is last in line. K.CC.4a

9. How many hats are there? Circle the right number. K.CC.3

14 15 16

10. Write the number that makes ten. K.OA.4

3 and 7

11. Write the number that matches the number of horns shown. K.CC.4a, K.CC.3

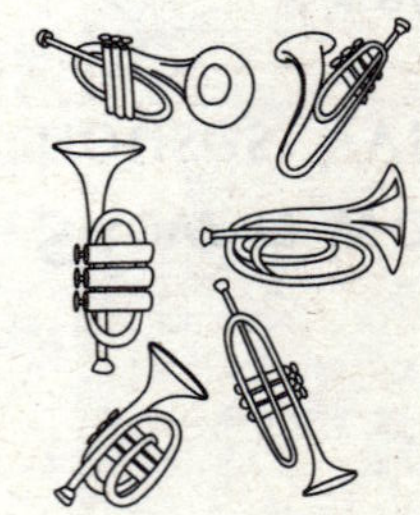

6

12. Circle the group with more bees. K.CC.6

13. Count the wagons. Write the number. Count the wheelbarrows. Write the number. Circle the group and number that is greater than the other. K.CC.3, K.CC.6

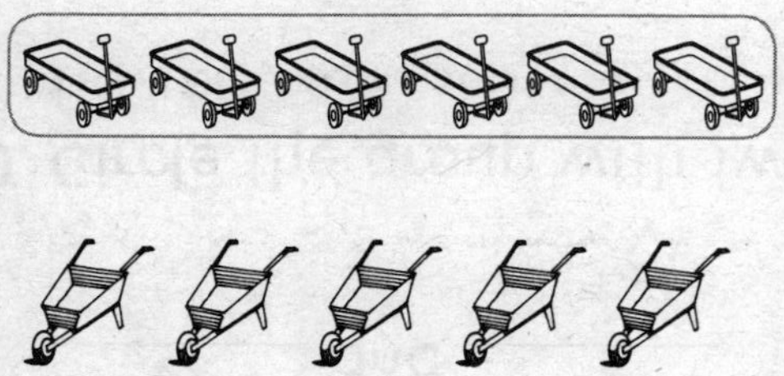

6

5

14. Color more boxes in the ten-frame to show sixteen. Write the number. K.CC.3

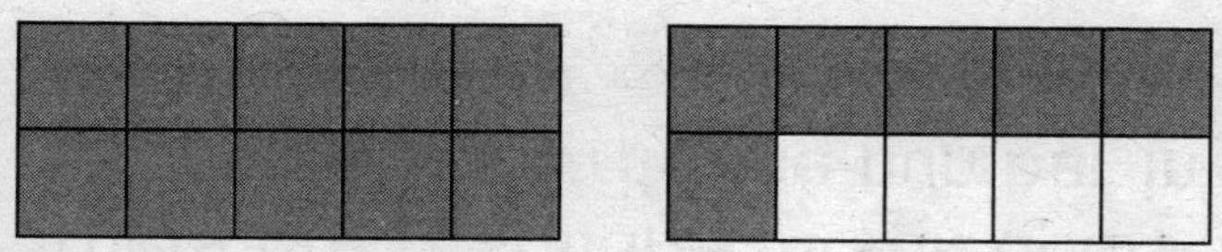

16

15. Match the picture to the statement. K.OA.3

4 and 1 is 5

4 and 5 is 9

16. Shade the ten frame to make eight. Write the number. K.CC.4

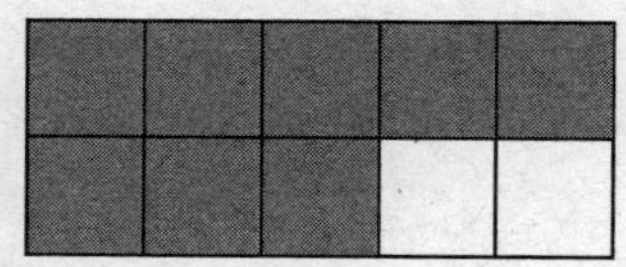

8

17. Write the number of birds. K.CC.3, K.CC.5

4

Copyright © McGraw-Hill Education. Permission is granted to reproduce for classroom use.

Copyright © McGraw-Hill Education. Permission is granted to reproduce for classroom use.

18. Circle the number that is one more than 12. K.CC.1

13 14 15

19. There are 12 squares. Break apart a group of 8. Write the number in the other group. K.OA.3

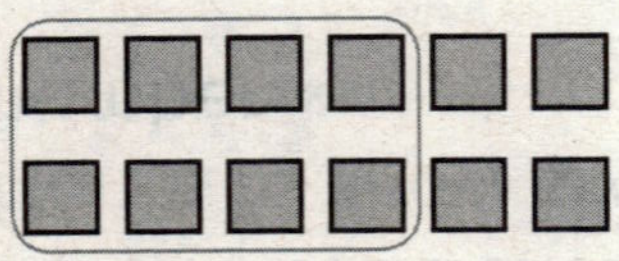

8 and 4

20. Circle the group with two flowers. K.CC.4a, K.CC.1

Page 156 • Building Pianos

Task Scenario Students will use counting and writing numbers up through 100 to answer questions about pianos. Students are required to research the number of keys on a piano.		
CCSS Content Standard(s)		K.CC.1, K.CC.4
Mathematical Practices		MP1, MP2, MP4
Depth of Knowledge		DOK2, DOK3
Part	**Maximum Points**	**Scoring Rubric**
A	2	Full Credit: 88 Partial Credit (1 point) will be given for the correct answer (88) without the number being formed correctly. No credit will be given for an incorrect answer.
B	3	Full Credit: 89, 90, 91 Partial Credit (1 point) will be given for each correct answer. No credit will be given for three incorrect answers.
C	2	Full Credit: 14 Partial Credit (1 point) will be given for the correct answer (14) without the number being formed correctly. No credit will be given for an incorrect answer.
TOTAL	7	

Performance Task Rubrics

NAME ______________ DATE ______________

SCORE ______________

Benchmark Test 2

1. There are two pigs in a pen. Three more pigs come in the pen. How many pigs are there in all? K.OA.1

5

2. Jaylan saw 6 bugs. Two crawled away. Cross out two bugs. Write how many are left. K.OA.1

4

3. Circle the number that shows "10 and 2 more." K.NBT.1

2 10 11 (12)

4. Count the numbers in each ten frame. Write the numbers. Write how many in all. K.NBT.1

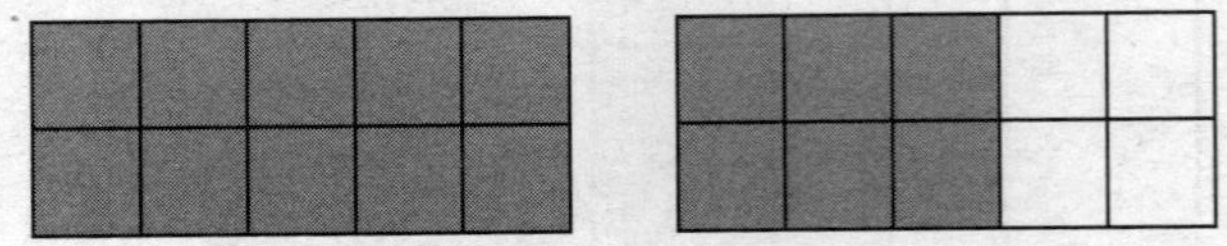

10 and 6 is 16

5. Count the number in each group. Write how many in all. K.OA.1

4 + 3 = 7 in all

6. Circle the pies to show K.OA.1

2 + 6 = 8

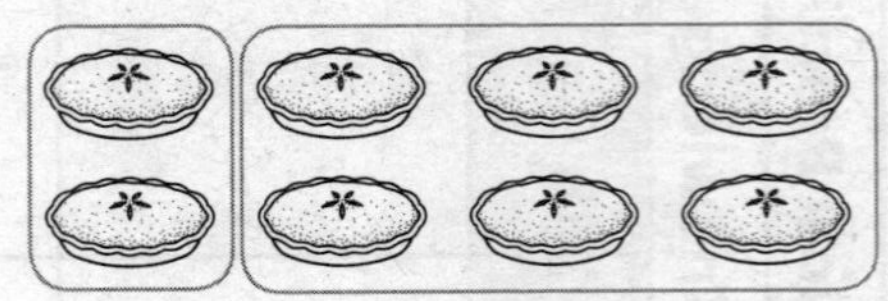

7. Count the number of each boat. Write each number. K.NBT.1

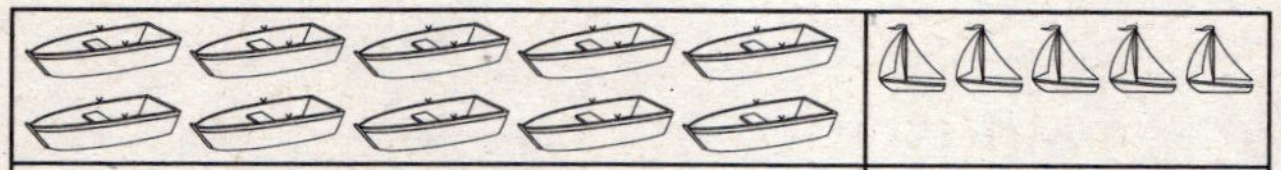

10 5

8. Compare the objects. Circle the one that is shorter. Underline the one that is longer. K.MD.2

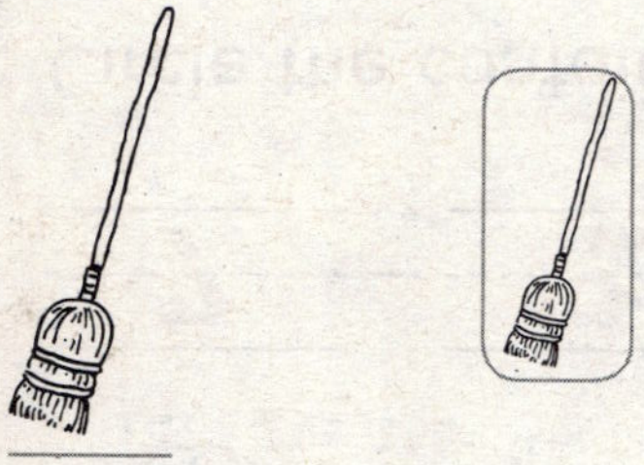

9. Underline the plus sign. Circle the subtraction sign. K.OA.1

$\times$ $-$ $+$ $=$

10. There are ten bikes. Six bikes ride away. Cross out six bikes. Write how many bikes are left. K.OA.3

4

11. Shade the ten frames to show "10 and 8 is 18." K.NBT.1

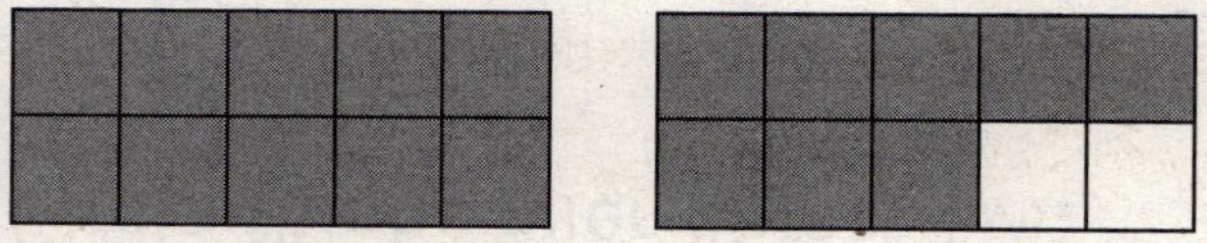

12. Circle the things that are too heavy to lift. K.MD.2

13. Count the number of bees. Count how many are away from the hive. Count how many are left in the hive. Write the numbers. K.OA.1

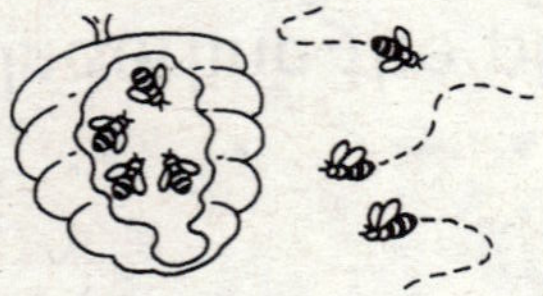

7 – 3 = 4

14. Circle the container that holds the most. K.MD.2

15. Henry has 4 balls. He loses one ball. How many balls does Henry have left? Circle the number. K.OA.1

1 2 3

16. Circle the heavier object. K.MD.2

17. There are 3 dry dogs and 3 wet dogs. How many dogs are there in all? Circle the correct number. K.OA.1

5 6 7

Copyright © McGraw-Hill Education. Permission is granted to reproduce for classroom use.

Copyright © McGraw-Hill Education. Permission is granted to reproduce for classroom use.

18. How many pennies long is the dart? K.MD.2

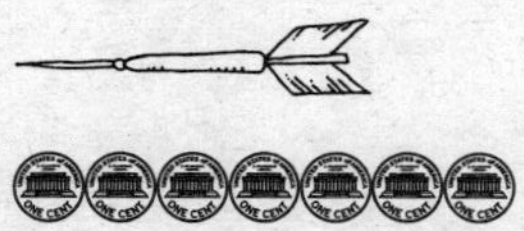

5

19. Cross out blocks to show the number sentence. K.OA.1

$10 - 2 = 8$

20. There are thirteen books on a shelf. Three books fell off the shelf. Cross out three books. How many books are left on the shelf? K.NBT.1

10

Page 164 • Studying Bugs

Task Scenario Students will use addition to count the number of bugs and legs on bugs. Students are required to research the number of legs on a spider and the number of legs on an insect.	
CCSS Content Standard(s)	K.CC.1, K.CC.3, K.OA.1
Mathematical Practices	MP1, MP2, MP4
Depth of Knowledge	DOK1, DOK2, DOK3

Part	Maximum Points	Scoring Rubric
A	3	Full Credit: Student draws a spider with 8 legs. Student draws an insect with 6 legs. 14 legs total Partial Credit (1 point) will be given for each of the three correct answers. No credit will be given for three incorrect responses.
B	3	Full Credit: $3 + 2 = 5$ Partial Credit (1 point) will be given for a correct answer for each of the entries in the number sentence (3, 2, and 5). No credit will be given for three incorrect numbers.
TOTAL	6	

NAME ______________________ DATE ______________

SCORE ______________

Benchmark Test 3

1. Write the number that is one more than 4. K.CC.2, K.CC.4c

5

2. Count the apples. Write the number. Count the bananas. Write the number. Circle the number that is greater than the other. K.CC.5, K.CC.7

7

9

3. Circle the shapes you can roll. K.G.B.4

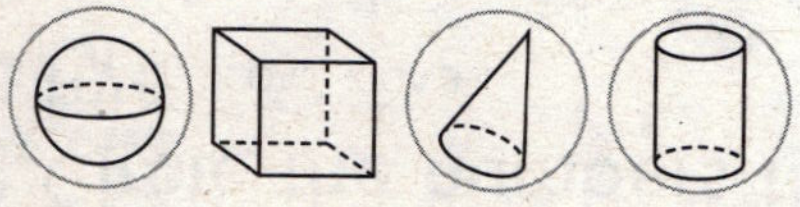

4. Color the boxes in the ten-frames to show 14. K.CC.4, K.CC.5

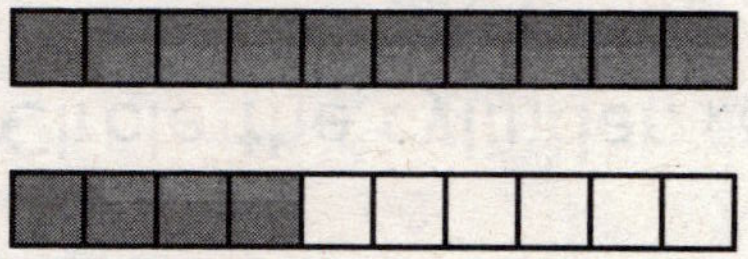

5. Count the tennis balls. How many are there? Write the number. K.CC.3, K.CC.5

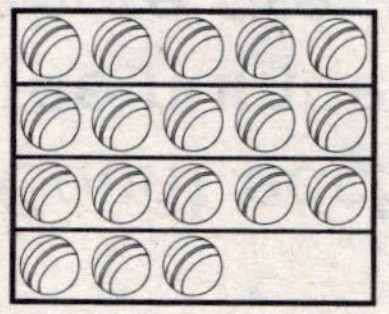

18

6. Count by 10s. Write the numbers. K.CC.1

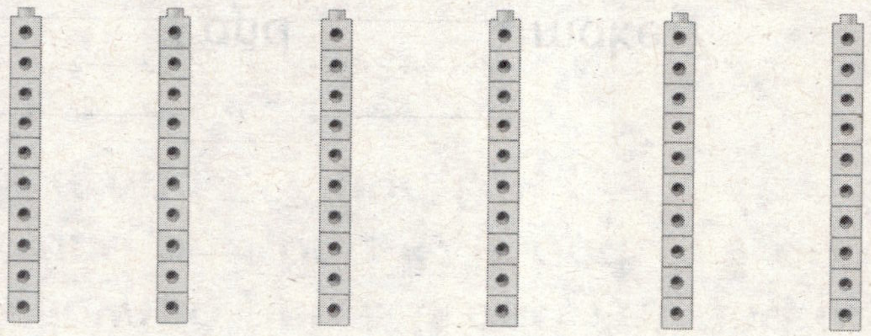

10 20 30 40 50 60

7. Circle the numbers that make five. K.OA.1, K.OA.3

2 and 3 | 2 and 4

1 and 4 | 1 and 3

8. What number does 3 and 1 make? Draw objects to show the number. Write the number. K.OA.2, K.OA.5

Students should draw four objects.

4

9. Look at the first shape in the row. Say the name of the shape to yourself. Compare this shape to the other shapes. Circle the matching shape. K.G.2, K.G.4

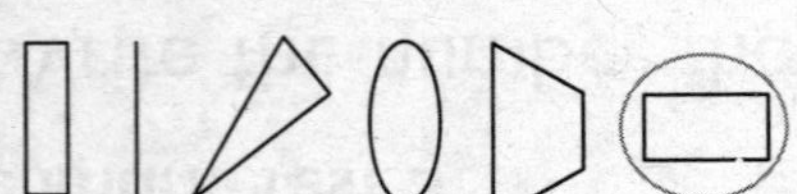

10. Write one way to make seven. Write another way to make seven. K.OA.1, K.OA.3

Students should write two of the following pairs: 0 and 7, 1 and 6, 2 and 5, 3 and 4, 4 and 3, 5 and 2, 6 and 1, 7 and 0

_____ and _____ make 7

_____ and _____ make 7

11. Count each group of crayons. Write the numbers. Write how many crayons there are in all. K.OA.1, K.OA.2

5 and 4 make 9

12. Circle the cylinder. K.G.4

13. Circle eight cars. Write the number of cars not circled to make 10. K.OA.4

8 and __2__ make 10

14. Draw two groups of fish to make four fish in all. Write the number in each group. K.OA.1, K.OA.3

Students should draw groups of 1 and 3 fish, 2 and 2 fish, or 3 and 1 fish, and write the matching numbers.

15. Draw an X through the number of hats that shows how many you would take away. K.OA.2

6 − 2 = 4

16. Match each word to the correct symbol. K.OA.1

equal sign	+
plus sign	−
minus sign	=

17. Eight apples are in the tree. You pick three. Draw an X on each apple you picked. How many are left? Write the number. K.OA.1, K.OA.2

__5__

18. Circle the group of objects that shows ten. Circle the group that shows some more. Write the number in each group. Write the number made. K.CC.4, K.NBT.1

10 and 3 make 13

19. Draw objects in the ten-frames to make sixteen. K.OA.1, K.NBT.1

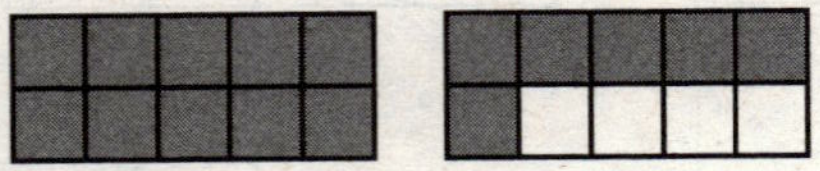

20. Count the number of objects in each ten-frame. Write each number. Trace the number made. K.CC.4, K.NBT.1

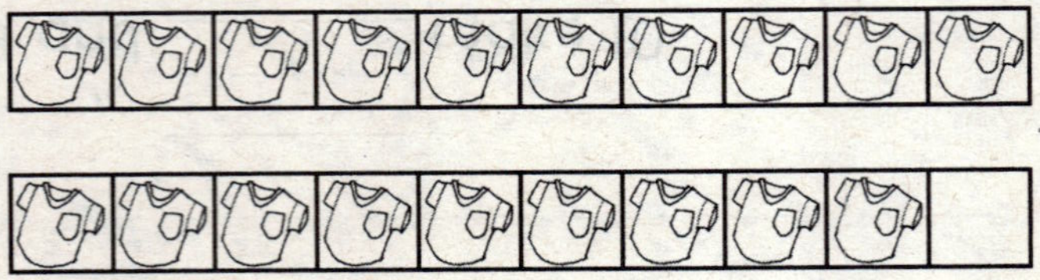

10 and 9 more make 19

21. Circle the train that is shorter. Draw an X through the train that is longer. K.MD.1, K.MD.2

22. Circle the animal that is lighter. Draw an X through the animal that is heavier. K.MD.1, K.MD.2

23. Circle the container that holds more. Draw an X through the container that holds less. K.MD.1, K.MD.2

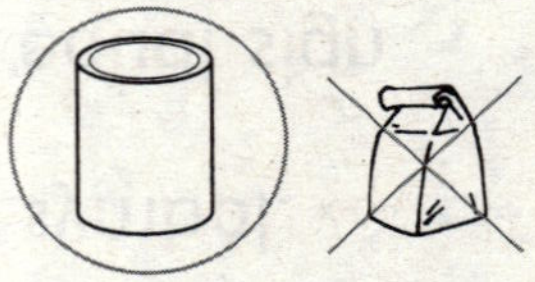

24. Circle the dog that is different. Draw an X through the dogs that are the same. K.MD.3

25. Circle the shape that is different. K.MD.3

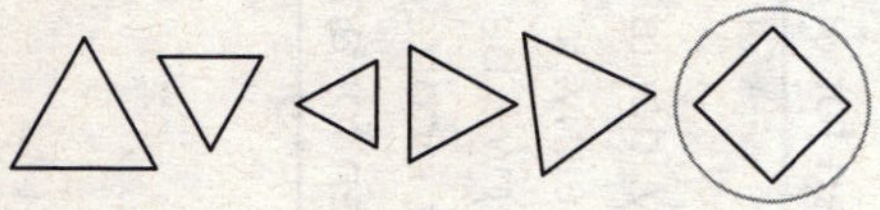

26. Circle each apple. Draw an X through each banana. Draw a box around each pear. K.MD.3

27. Circle the two-dimensional shapes. Draw an X through the three-dimensional shapes. K.G.B.3

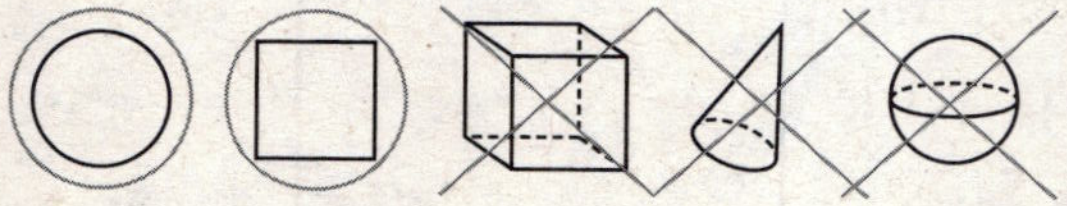

28. Draw a balloon above the table. Draw a box under the table. K.G.1

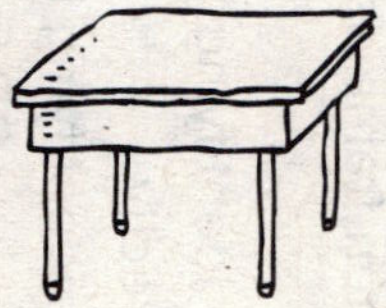

Students should draw a balloon above the table and a box below the table.

29. Circle the object beside the flower. K.G.1

30. Draw a triangle. K.G.5

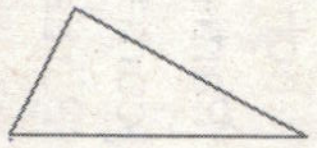

31. Circle the shape that has four sides. K.G.4

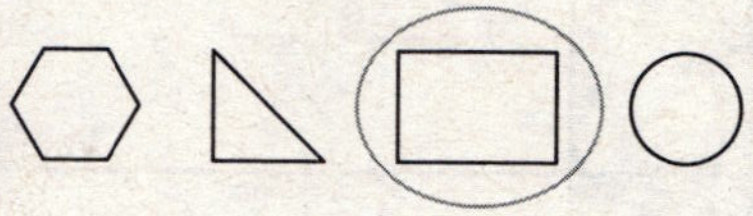

Page 175 • Packing Tomatoes

Task Scenario Students will count objects, write the associated number, represent addition and subtraction with drawings, and decompose numbers.		
CCSS Content Standard(s)		K.CC.3, K.CC.4b, K.CC.5, K.OA.1, K.OA.2, K.OA.4, K.NBT.1
Mathematical Practices		MP4, MP7
Depth of Knowledge		DOK1, DOK2, DOK3
Part	**Maximum Points**	**Scoring Rubric**
A	4	Full Credit: The student correctly identifies the quantity of tomatoes in the first ten-frame as 10 and in the second ten-frame as 2 and writes each number clearly. Partial Credit (3 points) will be given for identifying both quantities correctly, but writing one number incorrectly. Partial Credit (2 points) will be given for identifying both quantities correctly, but writing the numbers incorrectly **OR** for correctly identifying one quantity and writing the number correctly, but incorrectly identifying the other quantity. Partial Credit (1 point) will be given for correctly identifying one quantity, but incorrectly identifying the other quantity, and incorrectly writing both numbers. No credit will be given for an incorrect answer.
B	2	Full Credit: The student correctly identifies the quantity as 12 and writes the number clearly. Partial Credit (1 point) will be given for identifying the correct number but writing it incorrectly. No credit will be given for an incorrect answer.
C	2	Full Credit: The student correctly identifies the quantity needed as 8 and writes the number clearly. Partial Credit (1 point) will be given for identifying the correct number but writing it incorrectly. No credit will be given for an incorrect answer.

Part	Maximum Points	Scoring Rubric
D	3	Full Credit: The student correctly crosses out two tomatoes, identifies the remaining quantity as 3, and writes 3 correctly. Partial Credit (2 points) will be given for correctly crossing out two tomatoes, identifying the remaining quantity as 3, but writing 3 incorrectly **OR** for identifying the remaining quantity as 3, and writing 3 correctly, but crossing out the wrong number of tomatoes. Partial Credit (1 point) will be given for identifying the remaining quantity as 3, but writing 3 incorrectly and crossing out the wrong number of tomatoes. No credit will be given for an incorrect answer.
TOTAL	11	

NAME ____________ DATE ____________

Benchmark Test 4

SCORE ____________

1. Look at the first shape in the row. Say the name of the shape to yourself. Compare this shape to the other shapes. Circle the matching shape. K.G.2, K.G.4

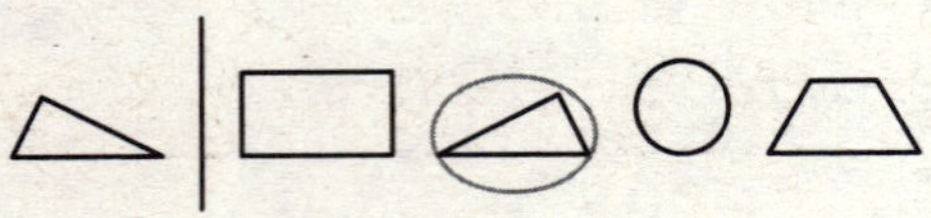

2. Circle the shovel that is shorter. Draw and X through the shovel that is longer. K.MD.1, K.MD.2

3. Circle the cupcake that is different. Draw an X through the cupcakes that are the same. K.MD.3

4. Write one way to make eight. Write another way to make eight. K.OA.1, K.OA.3

Sample answer:

6 and 2 make 8

5 and 3 make 8

5. What do 5 and 1 make? Draw circles to show the number. Write the number. K.OA.2, K.OA.5

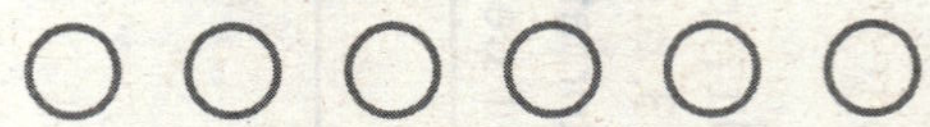

6

6. Count each group of candles. Write the numbers. Write how many candles there are in all. K.OA.1, K.OA.2

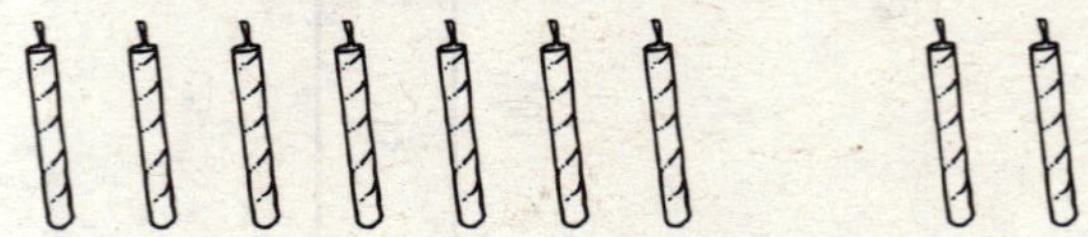

7 and 2 make 9

7. Draw objects in the ten-frame to show 14. K.OA.1, K.NBT.1

8. Circle the group of objects that shows ten. Circle the group that shows some more. Write the number in each group. Write the number made. K.CC.4, K.NBT.1

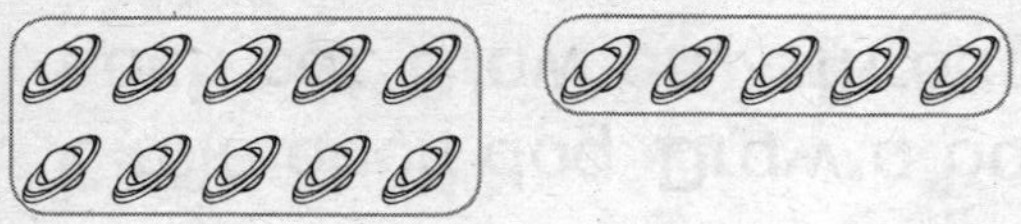

10 and 5 make 15

9. Count the cars. Write the number. Count the airplanes. Write the number. Circle the number that is greater than the other. K.CC.5, K.CC.7

8

6

10. Circle the numbers that make 5. K.OA.1, K.OA.3

0 and 5 4 and 2

1 and 5 2 and 3

11. Nine desks are in a classroom. The teacher removes three. Draw an X on each desk that is removed. How many are left? Write the number. K.OA.1, K.OA.2

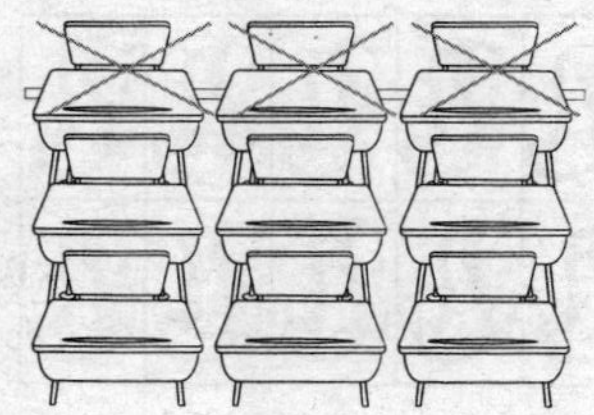

6

12. Circle the animal that is lighter. Draw an X through the animal that is heavier. K.MD.1, K.MD.2

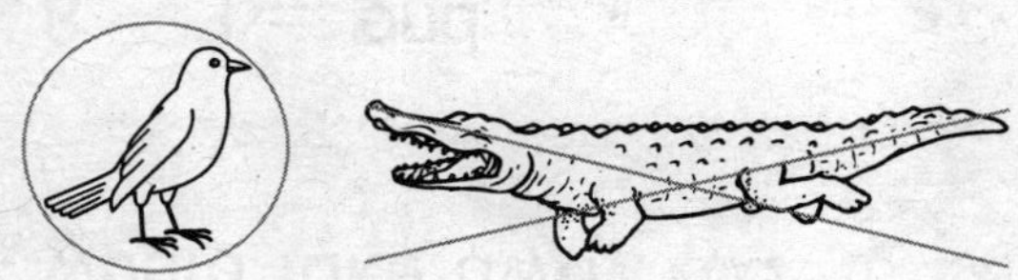

13. Circle the container that holds more. Draw an X through the container that holds less. K.MD.1, K.MD.2

14. Write the number that is one more than 3. K.CC.2, K.CC.4c

4

15. Circle each dog. Draw a box around each cat. Draw an X through each bird. K.MD.3

16. Draw an X through the number of sleds that shows how many you would take away. K.OA.2

8 − 3 = 5

17. Count the number of objects in each ten-frame. Write each number. Write the number made. K.CC.4, K.NBT.1

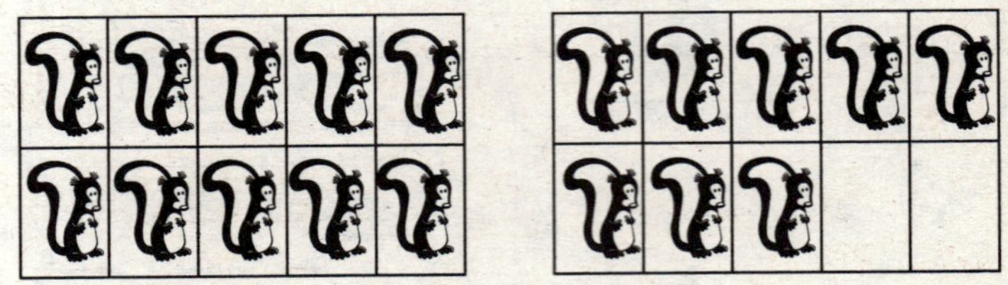

10 and 8 make 18

18. Circle the shape that has three sides. K.G.4

Copyright © McGraw-Hill Education. Permission is granted to reproduce for classroom use.

Copyright © McGraw-Hill Education. Permission is granted to reproduce for classroom use.

19. Count the baseball bats. How many are there? Write the number. K.CC.4, K.CC.5

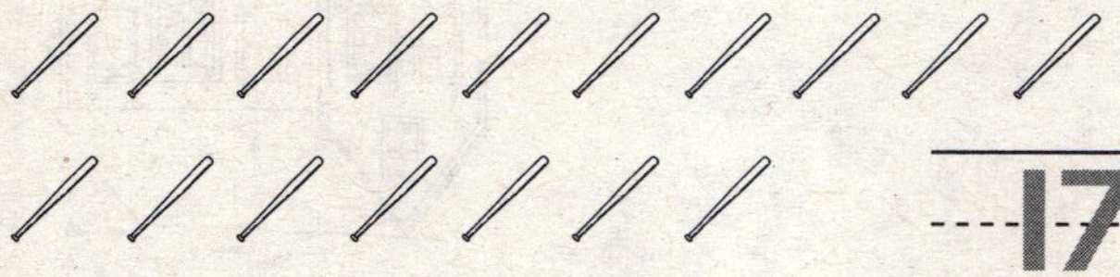

17

20. Count by 10s. Write the numbers. K.CC.1

10

20

30

40

21. Match each word to the correct symbol. K.OA.1

equal sign —
plus sign +
minus sign =

22. Circle the shape that is different. K.MD.3

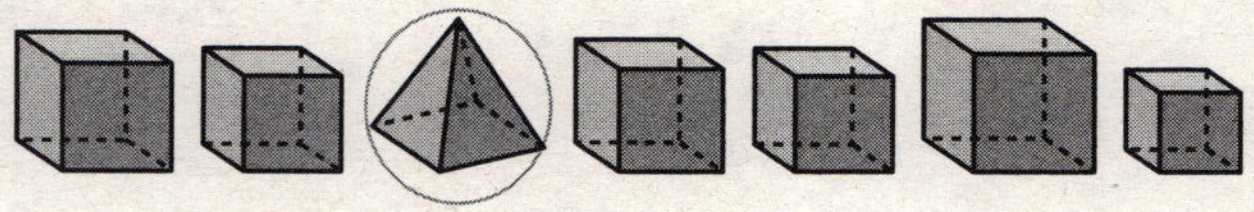

23. Draw two groups of smiley faces to make five smiley faces in all. Write the number in each group. K.OA.1, K.OA.3

Sample answer:

2 and 3

24. Circle the three-dimensional shapes. Draw an X through the two-dimensional shapes. K.G.3

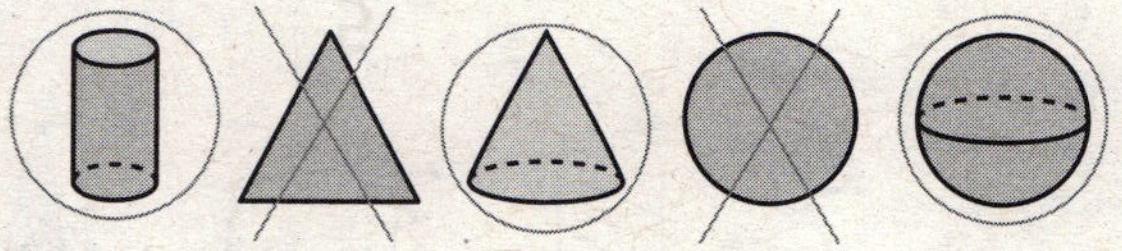

Copyright © McGraw-Hill Education. Permission is granted to reproduce for classroom use.

25. Circle the shapes you can stack. K.G.4

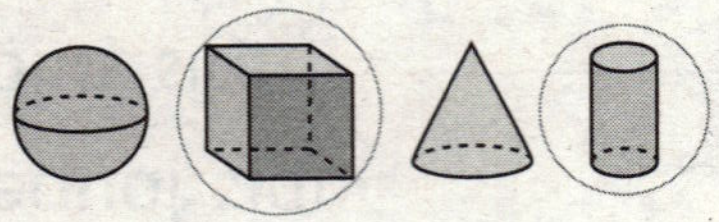

26. Circle seven beds. Write the number of beds not circled to make 10. K.OA.4

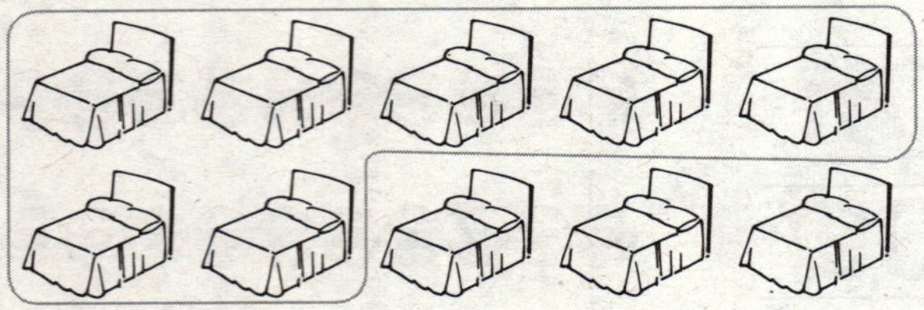

7 and 3 make 10

27. Draw a bird above the house. Draw a worm below the house. K.G.1

28. Circle the cone. K.G.4

29. Color the boxes in the ten-frames to show 13. K.CC.3, K.CC.5

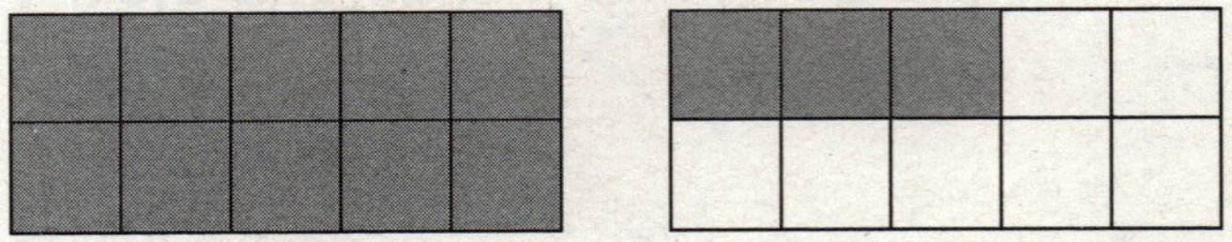

30. Circle the object beside the hammer. K.G.1

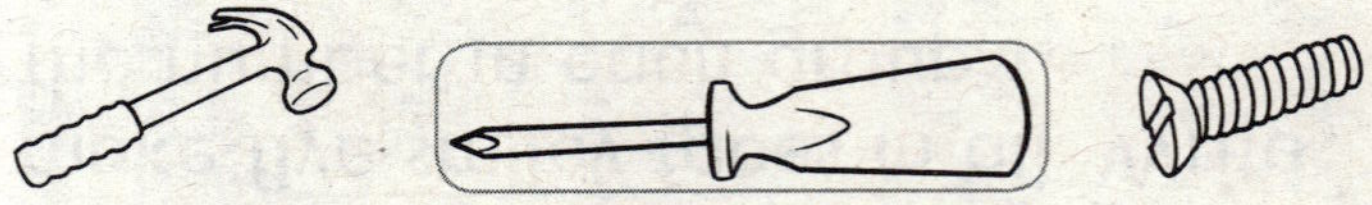

31. Draw a square. K.G.5

Page 187 • Stacking Books

Task Scenario	
Students will count objects, write the associated number, represent addition and subtraction with drawings, and decompose numbers.	
CCSS Content Standard(s)	K.CC.3, K.CC.4b, K.CC.5, K.OA.1, K.OA.2, K.OA.4, K.NBT.1
Mathematical Practices	MP4, MP7
Depth of Knowledge	DOK1, DOK2, DOK3

Part	Maximum Points	Scoring Rubric
A	4	Full Credit: The student correctly identifies the quantity of books in the first ten-frame as 10 and in the second ten-frame as 3 and writes each number clearly. Partial Credit (3 points) will be given for identifying both quantities correctly, but writing one number incorrectly. Partial Credit (2 points) will be given for identifying both quantities correctly, but writing the numbers incorrectly **OR** for correctly identifying one quantity and writing the number correctly, but incorrectly identifying the other quantity. Partial Credit (1 point) will be given for correctly identifying one quantity, but incorrectly identifying the other quantity, and incorrectly writing both numbers. No credit will be given for an incorrect answer.
B	2	Full Credit: The student correctly identifies the quantity as 13 and writes the number clearly. Partial Credit (1 point) will be given for identifying the correct number but writing it incorrectly. No credit will be given for an incorrect answer.
C	2	Full Credit: The student correctly identifies the quantity needed as 7 and writes the number clearly. Partial Credit (1 point) will be given for identifying the correct number but writing it incorrectly. No credit will be given for an incorrect answer.

Part	Maximum Points	Scoring Rubric
D	3	Full Credit: The student correctly crosses out two books, identifies the remaining quantity as 4, and writes 4 correctly. Partial Credit (2 points) will be given for correctly crossing out two books, identifying the remaining quantity as 4, but writing 4 incorrectly **OR** for identifying the remaining quantity as 4, and writing 4 correctly, but crossing out the wrong number of books. Partial Credit (1 point) will be given for identifying the remaining quantity as 4, but writing 4 incorrectly and crossing out the wrong number of books. No credit will be given for an incorrect answer.
TOTAL	11	